REALIZING THE POTENTIAL OF C4I

FUNDAMENTAL CHALLENGES

Committee to Review DOD C4I Plans and Programs

Computer Science and Telecommunications Board

Commission on Physical Sciences, Mathematics, and Applications

National Research Council

NATIONAL ACADEMY PRESS
Washington, D.C.   1999

NOTICE: The project that is the subject of this report was approved by the Governing Board of the National Research Council, whose members are drawn from the councils of the National Academy of Sciences, the National Academy of Engineering, and the Institute of Medicine. The members of the committee responsible for the report were chosen for their special competences and with regard for appropriate balance.

Support for this project was provided by the Department of Defense. Any opinions, findings, conclusions, or recommendations expressed in this material are those of the authors and do not necessarily reflect the views of the sponsor.

Library of Congress Catalog Card Number 99-62272
International Standard Book Number 0-309-06485-6

Additional copies of this report are available from:
National Academy Press
2101 Constitution Avenue, NW
Box 285
Washington, DC 20055
800/624-6242
202/334-3313 (in the Washington Metropolitan Area)
http://www.nap.edu

Copyright 1999 by the National Academy of Sciences. All rights reserved.

Printed in the United States of America

COMMITTEE TO REVIEW DOD C4I PLANS AND PROGRAMS

JAMES C. McGRODDY, IBM (retired), *Chair*
CHARLES HERZFELD, Private Consultant, *Vice Chair*
NORMAN ABRAMSON, ALOHA Networks
EDWARD BALKOVICH, Bell Atlantic
JORDAN BARUCH, Jordan Baruch Associates
RICHARD BASEIL, Telcordia Technologies (formerly Bellcore)
THOMAS A. BERSON, Anagram Laboratories
RICHARD KEMMERER, University of California at Santa Barbara
BUTLER LAMPSON, Microsoft Corporation
DAVID M. MADDOX, Gen USA (retired), Private Consultant
PAUL D. MILLER, ADM USN (retired), Alliant Technology Systems
CARL G. O'BERRY, Lt Gen USAF (retired)
JOHN H. QUILTY, MITRE Corporation
ROBERT H. REED, Gen USAF (retired), Lear Astronics Corporation
H. GREGORY TORNATORE, Applied Physics Laboratory, Johns Hopkins University

Staff

HERBERT S. LIN, Senior Scientist and Study Director
JON EISENBERG, Program Officer
JULIE ESANU, Research Associate
MARK BALKOVICH, Research Associate
MICKELLE RODGERS, Project Assistant
NICCI T. DOWD, Project Assistant
DAVID PADGHAM, Project Assistant

COMPUTER SCIENCE AND TELECOMMUNICATIONS BOARD

DAVID D. CLARK, Massachusetts Institute of Technology, *Chair*
FRANCES E. ALLEN, IBM T.J. Watson Research Center
JAMES CHIDDIX, Time Warner Cable
JOHN M. CIOFFI, Stanford University
W. BRUCE CROFT, University of Massachusetts at Amherst
A.G. (SANDY) FRASER, AT&T
SUSAN L. GRAHAM, University of California at Berkeley
JAMES GRAY, Microsoft Corporation
PATRICK M. HANRAHAN, Stanford University
JUDITH HEMPEL, University of California at San Francisco
BUTLER W. LAMPSON, Microsoft Corporation
EDWARD D. LAZOWSKA, University of Washington
DAVID LIDDLE, Interval Research
JOHN MAJOR, Wireless Knowledge
TOM M. MITCHELL, Carnegie Mellon University
DONALD NORMAN, Nielsen Norman Group
RAYMOND OZZIE, Groove Networks
DAVID A. PATTERSON, University of California at Berkeley
LEE SPROULL, Boston University
LESLIE L. VADASZ, Intel Corporation

Staff

MARJORY S. BLUMENTHAL, Director
HERBERT S. LIN, Senior Scientist
JERRY R. SHEEHAN, Senior Program Officer
ALAN S. INOUYE, Program Officer
JON EISENBERG, Program Officer
JANET D. BRISCOE, Administrative Associate
RITA GASKINS, Project Assistant
NICCI T. DOWD, Project Assistant
DAVID PADGHAM, Project Assistant
MARGARET MARSH, Project Assistant

COMMISSION ON PHYSICAL SCIENCES, MATHEMATICS, AND APPLICATIONS

PETER M. BANKS, ERIM International, Inc., *Co-chair*
W. CARL LINEBERGER, University of Colorado, *Co-chair*
WILLIAM BROWDER, Princeton University
LAWRENCE D. BROWN, University of Pennsylvania
MARSHALL H. COHEN, California Institute of Technology
RONALD G. DOUGLAS, Texas A&M University
JOHN E. ESTES, University of California at Santa Barbara
JERRY P. GOLLUB, Haverford College
MARTHA P. HAYNES, Cornell University
JOHN L. HENNESSY, Stanford University
CAROL M. JANTZEN, Westinghouse Savannah River Company
PAUL G. KAMINSKI, Technovation, Inc.
KENNETH H. KELLER, University of Minnesota
MARGARET G. KIVELSON, University of California at Los Angeles
DANIEL KLEPPNER, Massachusetts Institute of Technology
JOHN KREICK, Sanders, a Lockheed Martin Company
MARSHA I. LESTER, University of Pennsylvania
M. ELISABETH PATÉ-CORNELL, Stanford University
NICHOLAS P. SAMIOS, Brookhaven National Laboratory
CHANG-LIN TIEN, University of California at Berkeley

NORMAN METZGER, Executive Director

The National Academy of Sciences is a private, nonprofit, self-perpetuating society of distinguished scholars engaged in scientific and engineering research, dedicated to the furtherance of science and technology and to their use for the general welfare. Upon the authority of the charter granted to it by the Congress in 1863, the Academy has a mandate that requires it to advise the federal government on scientific and technical matters. Dr. Bruce Alberts is president of the National Academy of Sciences.

The National Academy of Engineering was established in 1964, under the charter of the National Academy of Sciences, as a parallel organization of outstanding engineers. It is autonomous in its administration and in the selection of its members, sharing with the National Academy of Sciences the responsibility for advising the federal government. The National Academy of Engineering also sponsors engineering programs aimed at meeting national needs, encourages education and research, and recognizes the superior achievements of engineers. Dr. William A. Wulf is president of the National Academy of Engineering.

The Institute of Medicine was established in 1970 by the National Academy of Sciences to secure the services of eminent members of appropriate professions in the examination of policy matters pertaining to the health of the public. The Institute acts under the responsibility given to the National Academy of Sciences by its congressional charter to be an adviser to the federal government and, upon its own initiative, to identify issues of medical care, research, and education. Dr. Kenneth I. Shine is president of the Institute of Medicine.

The National Research Council was organized by the National Academy of Sciences in 1916 to associate the broad community of science and technology with the Academy's purposes of furthering knowledge and advising the federal government. Functioning in accordance with general policies determined by the Academy, the Council has become the principal operating agency of both the National Academy of Sciences and the National Academy of Engineering in providing services to the government, the public, and the scientific and engineering communities. The Council is administered jointly by both Academies and the Institute of Medicine. Dr. Bruce Alberts and Dr. William A. Wulf are chairman and vice chairman, respectively, of the National Research Council.

Preface

The Defense Authorization Act for Fiscal Year 1996 requested that the National Research Council (NRC) undertake a review of current and planned service and defense-wide programs for command, control, communications, computers, and intelligence (C4I) with a special focus on cross-service and inter-service issues (Box P.1). Programs for C4I account for some of the most complex systems, technologies, and functions in the military. Expenditures on C4I represent a significant fraction of the defense budget. C4I programs provide an interrelated group of capabilities that are distributed horizontally across the military services and vertically within each defense function.

Recognizing the potential leverage that enhanced C4I capabilities can provide to the various defense functions (e.g., battlespace situational awareness at all levels of the military command structure, tactical communications, target identification and acquisition, logistics, and so on), the Department of Defense (DOD) has begun major efforts to integrate the various C4I systems into a system of systems. This vision of a defense-wide rationalization of C4I architectures and systems—articulated in *Joint Vision 2010*[1]—is highly ambitious, and one that will undoubtedly stress traditional DOD ways of doing business.

In response to the legislative mandate, the Computer Science and Telecommunications Board (CSTB) of the NRC formed the Committee to

---

[1] Chairman of the Joint Chiefs of Staff. 1996. *Joint Vision 2010*, Joint Chiefs of Staff, Washington, D.C.

> **BOX P.1 Legislative Charge to the National Research Council**
>
> **Public Law 104-106**
> **Defense Authorization Act for Fiscal Year 1996**
>
> SEC. 262. REVIEW OF C4I BY NATIONAL RESEARCH COUNCIL.
>
> (a) Review by National Research Council—Not later than 90 days after the date of the enactment of this Act, the Secretary of Defense shall enter into a contract with the National Research Council of the National Academy of Sciences to conduct a comprehensive review of current and planned service and defense-wide programs for command, control, communications, computers, and intelligence (C4I) with a special focus on cross-service and inter-service issues.
>
> (b) Matters To Be Assessed in Review—The review shall address the following:
>
> (1) The match between the capabilities provided by current service and defense-wide C4I programs and the actual needs of users of these programs.
>
> (2) The interoperability of service and defense-wide C4I systems that are planned to be operational in the future.
>
> (3) The need for an overall defense-wide architecture for C4I.
>
> (4) Proposed strategies for ensuring that future C4I acquisitions are compatible and interoperable with an overall architecture.
>
> (5) Technological and administrative aspects of the C4I modernization effort to determine the soundness of the underlying plan and the extent to which it is consistent with concepts for joint military operations in the future.
>
> (c) Two-Year Period for Conducting Review—The National Research Council shall conduct the review over the two-year period beginning upon completion of the performance of the contract described in subsection (a).
>
> (d) Reports—(1) The National Research Council shall submit to the Department of Defense and Congress interim reports and progress updates on a regular basis as the review proceeds. A final report on the review shall set forth the findings, conclusions, and recommendations of the Council for defense-wide and service C4I programs and shall be submitted to the Committee on Armed Services of the Senate, the Committee on National Security of the House of Representatives, and the Secretary of Defense.
>
> (2) To the maximum degree possible, the final report shall be submitted in unclassified form with classified annexes as necessary.
>
> (e) Interagency Cooperation With Study—All military departments, defense agencies, and other components of the Department of Defense shall cooperate fully with the National Research Council in its activities in carrying out the review under this section.
>
> (f) Expedited Processing of Security Clearances for Study—For the purpose of facilitating the commencement of the study under this section, the Secretary of Defense shall expedite to the fullest degree possible the processing of security clearances that are necessary for the National Research Council to conduct the study.
>
> (g) Funding—Of the amount authorized to be appropriated in section 201 for defense-wide activities, $900,000 shall be available for the study under this section.

Review DOD C4I Plans and Programs. Unlike the groups responsible for many other studies of C4I (some of which are listed in Appendix B), the membership of the committee was about evenly divided between information technology experts from the commercial, non-defense sector and individuals with significant experience in military operations, either as senior commanders or as defense technologists (Appendix C). The motivation for this balance was that the civilian experts would bring perspective and insights from the commercial world that are relevant to the DOD, while the military experts would provide context and a sense of history and operational experience on what is, after all, a unique organization with a unique mission.

In the course of its work, the committee received briefings from DOD officials and others and conducted a number of site visits (Appendix A provides a list), reviewed recent reports described in Appendix B, and met seven times to discuss the input from these sources as well as the independent observations and findings drawn from the experience of the committee members themselves. With the limited resources and time available, the committee chose an approach in which it examined selected field exercises and various C4I programs, rather than attempting to conduct an exhaustive review of all C4I programs within DOD. The committee selected the particular programs and exercises it visited, although its DOD liaisons provided valuable input on possible subjects for examination. As a result of this approach, the findings and conclusions in this report are based on judgments resulting from the "on the ground" sense the committee developed through its sampling process, as well as the experience and knowledge of committee members. Sampling is by definition not comprehensive. However, the sampled data points are most likely to reflect the modal state of practice, and thus conclusions drawn on the basis of sampling are likely to be valid in some "average" sense.

One critical question faced by the committee during the course of its project was whether to interpret its legislative charge broadly or narrowly. The narrowest interpretation would have led to a detailed technical review of current DOD efforts in C4I architecture and standards to promote interoperability. The broadest interpretation would have led to an examination of all of the elements needed to sustain a revolution in military affairs based on C4I and information technology, including but not limited to technical considerations of interoperability. Taking the narrowest interpretation risked focusing only on the current state of affairs without taking into account future possibilities. Taking the broadest interpretation risked undertaking an assessment too large to be doable in any meaningful way given the time and resources available to the committee. After considerable discussion, the committee decided that a broad—but not the broadest—interpretation was appropriate. Interoperability would remain a central part of the report, but other issues as they emerged in the context of the interoperability problem would also be addressed.

Finally, it is appropriate to point out what this report does not cover:

- The report accepts as a given the proposition that C4I and information technologies will be central to the vision for the nation's fighting forces in the future. The committee recognizes some controversy over the validity of this proposition but, given the legislative charge, believes that it was a reasonable presumption. The committee believes that C4I has been and will continue to be a critical factor in any imaginable evolutionary path for U.S. forces; the argument in this area is one of degree rather than kind.
- The report focuses on three key areas—interoperability, information systems security, and DOD process and culture—that demand serious attention if the military promise of C4I is to be realized. Though issues in these areas were regarded as the most critical problems for DOD, this focus does not mean that other issues are not important as well. For example, issues related to data overload (a user of a C4I system being inundated with information that may be nice to have but is not critical) and data quality (knowing that the data entered into the system, such as a sensor input, is in fact valid) are acknowledged but not addressed explicitly, except as they interact with the critical issues of interoperability and information systems security.
- The report does not evaluate specific C4I programs or systems. While such evaluations are useful from a programmatic standpoint, the enormous number of C4I programs within DOD made such a task impossible within the context of this study. Furthermore, agencies such as the General Accounting Office are better suited to undertaking the analysis of specific programs. In general, the committee did look at some programs and did discuss technology issues, but did not conduct a programmatic audit as GAO might do, believing that its primary efforts would be better spent on examining the overall systemic challenges in exploiting the leverage of C4I.
- The report does not address the special issues related to the "online war" in which traditional media such as CNN and emerging online media such as Web sites and Internet relay chats help to influence public opinion, either directly or indirectly. This phenomenon is likely to become more important in the future as the United States conducts military operations in non-traditional venues, but the committee did not have the expertise to address these matters.
- The report does not address systems that are intended primarily to support strategic or national intelligence collection. The committee recognizes the fact that national intelligence assets may be used for tactical

purposes, and to the extent that this is true, the report's analysis, findings, and recommendations are relevant. Nevertheless, an examination of national intelligence programs was beyond the scope of the committee's primary focus.

The committee wishes to thank the Department of Defense for providing liaison and logistical support. In particular, Mr. John Buchheister and Mr. Jack Zavin (both from the Office of the Assistant Secretary of Defense for C3I) were enormously helpful and provided valuable guidance in helping to find appropriate briefers and information the committee needed for its study. The committee also wishes to acknowledge the efforts of those listed in Appendix A. Those responsible for hosting the committee on its site visits provided access and support, while those presenting briefings to the committee answered a wide variety of questions. Finally, committee members spoke to a number of personnel on its site visits who were not part of any formal presentation. Nevertheless, these personnel helped to provide the committee with a measure of "ground truth" and in general impressed the committee with their dedication to duty and their technical sophistication.

A special note of appreciation is due the NRC staff on this project. Without the insights and capable efforts of Dr. Herb Lin, it would have been impossible to assemble the team whose breadth of experience and knowledge were essential to the creation of this report. In addition, the committee would like to acknowledge Dr. Lin's expert support in developing an overall plan and specific agendas for its meetings and site visits. The NRC team of Lin and Dr. Jon Eisenberg, working together with Mr. Buchheister and Mr. Zavin, developed a plan of briefings with key people and effective site visits that enabled the committee to focus on the relevant issues and rapidly develop a base of common knowledge in the complex area of C4I. The committee received major and capable help and support from Lin and Eisenberg, in both developing an effective process for the writing effort and ensuring a consistent style and, more important, finding a reasonable consensus on key issues and recommendations without diluting the directness and forcefulness of the committee's sentiments. As always, the committee, in its collective authorship, is responsible for the analysis, findings, and recommendations of this report.

Acknowledgment of Reviewers

This report was reviewed by individuals chosen for their diverse perspectives and technical expertise, in accordance with procedures approved by the National Research Council's (NRC's) Report Review Committee. The purpose of this independent review is to provide candid and critical comments that will assist the authors and the NRC in making the published report as sound as possible and to ensure that the report meets institutional standards for objectivity, evidence, and responsiveness to the study charge. The contents of the review comments and draft manuscript remain confidential to protect the integrity of the deliberative process. We wish to thank the following individuals for their participation in the review of this report:

W. Earl Boebert, Sandia National Laboratories,
William Crowell, Cylink,
Gerald Dinneen, Honeywell, Inc. (retired),
Jack Donegan, San Diego Supercomputer Center,
Robert Everett, MITRE Corporation,
Louis Finch, STR,
Mike Frankel, SRI International,
Richard L. Hearney, The Boeing Company,
David Heebner, Private Consultant,
John C. Henderson, Boston University,
Anita Jones, University of Virginia,
Stephen Kent, BBN,
John B. LaPlante, Burdeshaw Associates, Limited,

Stephen Lukasik, Independent Consultant,
Larry Lynn, Private Consultant,
Stuart E. Madnick, Massachusetts Institute of Technology,
Noel D. Matchett, Information Security Incorporated,
James McCarthy, United States Air Force Academy,
Wesley L. McDonald, U.S. Navy (retired),
Robert Nesbit, MITRE Corporation,
Kumar Patel, University of California at Los Angeles,
Stu Personick, Bell Laboratories,
William H. Press, Harvard University,
Jeff Rulifson, Sun Microsystems,
Casmir S. Skrzypczak, Bellcore,
Harry Train II, SAIC,
Harry Van Trees, George Mason University,
Andrew Viterbi, QUALCOMM,
Willis H. Ware, RAND Corporation,
Larry D. Welch, Institute for Defense Analyses,
Albert D. Wheelon, Hughes Aircraft Company (retired),
Sheila Widnall, Massachusetts Institute of Technology,
Len Wishart, U.S. Army (retired), and
John Yeosock, U.S. Army (retired).

Although the individuals listed above provided many constructive comments and suggestions, responsibility for the final content of this report rests solely with the study committee and the NRC.

Contents

NOTE TO THE READER: The executive summary is an essential component of this report. In addition to collecting the findings and recommendations presented in Chapters 2 through 4, the executive summary distills the goals and principles described in the main text of the report as informing the effective use of C4I systems and technology for military operations. This distillation is not to be found elsewhere in the report.

EXECUTIVE SUMMARY		1
1	INTRODUCTION	27
	1.1 What Is C4I?	28
	1.2 The Impact of C4I on Military Effectiveness	30
	1.2.1 Evidence from Recent Experience	30
	1.2.2 Potential Impact of C4I on Military Operations	36
	1.3 The U.S. Military's Work in Exploiting Information Technology	43
	1.4 The Role of C4I Systems in Future Military Environments	43
	1.4.1 Likely Environments of Future Military Operations	43
	1.4.2 Rapid Planning to Support Rapid Response	50
	1.4.3 Support for Deployment of Forces in the Changing Environment	52

	1.4.4	Proliferation in the Use of the U.S. Military for Sustainment and Support Operations (Military Operations Other Than War)	53
	1.4.5	Complexities of Exercising Command and Control of Forces in Regional Conflict Environments	56
	1.4.6	Strategic Vulnerability of Infrastructure to Information Attack	56
1.5	Expected Information Technology Trends for C4I		57
	1.5.1	Computers	58
	1.5.2	Communications	58
	1.5.3	Sensors	59
	1.5.4	Weapons	59
1.6	DOD Budget and Organizational Structure for C4I		60
	1.6.1	Budget	60
	1.6.2	DOD Organizational Structure for C4I	61
1.7	Challenges to the Exploitation of the Military Leverage of C4I		62

2 INTEROPERABILITY 64

2.1	What Is Interoperability and Why Is It Important?		64
	2.1.1	What Is Interoperability?	64
	2.1.2	Why Interoperability Is Important	69
	2.1.3	Dimensions of Technical Interoperability	72
2.2	Why Achieving Interoperability Is Difficult		73
	2.2.1	Challenges Common to All Large Enterprises	73
	2.2.2	Special Challenges Faced by the Department of Defense	80
2.3	Technical Approaches to Interoperability		84
	2.3.1	Architecture	85
	2.3.2	Interfaces, Layers, and Middleware	85
	2.3.3	Standards	87
	2.3.4	Data Interoperability	90
	2.3.5	Developing and Implementing Architectures	92
2.4	Testing		96
2.5	DOD Interoperability Strategy		100
	2.5.1	Overview	100
	2.5.2	Elements of the DOD Strategy	101
2.6	Measuring Interoperability		105
	2.6.1	A Technical Compliance Scorecard	105
	2.6.2	A Systems Interoperability Scorecard	106
	2.6.3	An Operational (Mission-Enabling) Interoperability Scorecard	107

	2.7	Findings	109
	2.8	Recommendations	116
3	INFORMATION SYSTEMS SECURITY		130
	3.1	Introduction	130
		3.1.1 Vulnerabilities in Information Systems and Networks	135
		3.1.2 Security Requirements	136
		3.1.3 Role of Cryptography	137
	3.2	Major Challenges to Information Systems Security	139
		3.2.1 The Asymmetry Between Defense and Offense	139
		3.2.2 Networked Systems	139
		3.2.3 Ease-of-Use Compromises	140
		3.2.4 Perimeter Defense	140
		3.2.5 The Use of COTS Components	141
		3.2.6 Threats Posed by Insiders	142
		3.2.7 Passive Defense	143
	3.3	Defensive Functions	144
	3.4	Responsibility for Information Systems Security in DOD	152
	3.5	The Information Systems Security Threat	154
	3.6	Technical Assessment of C4I System Security	156
	3.7	Findings	157
	3.8	Recommendations	160
4	PROCESS AND CULTURE		179
	4.1	Managing Change	179
		4.1.1 Clear Vision for the Future	180
		4.1.2 Supporting Processes	180
		4.1.3 Persistent Leadership Creating a Sense of Urgency	182
		4.1.4 Process Reengineering	184
		4.1.5 Budgets and Reprioritization of Investment	186
	4.2	Special Non-Technical Challenges Faced by the Military	187
		4.2.1 Situational Challenges	187
		4.2.2 Organizational Challenges	188
		4.2.3 Schedule and Budget Challenges	192
		4.2.4 Coalition Challenges	193
	4.3	The Acquisition System	194
		4.3.1 Overview	194
		4.3.2 Requirements, the 80% Solution, and Functional Specifications	197

		4.3.3	Exploiting Commercial Technology	199
		4.3.4	Testing	204
		4.3.5	Flexibility in the Process	205
		4.3.6	Support of the Legacy Base Versus New Technology	207
	4.4	Personnel, Knowledge, and Professionalism		208
	4.5	Exercises, Experiments, and Doctrinal Change		210
	4.6	Management Metrics and Measures of Military Effectiveness		212
		4.6.1	DOD Use of Management Metrics and Measures of Military Effectiveness	213
		4.6.2	Considerations in Assessment of C4I System Effectiveness	216
		4.6.3	Caveats	217
		4.6.4	Ways of Generating and Developing Data	219
	4.7	Findings		221
	4.8	Recommendations		229

APPENDIXES

A	List of Site Visits and Briefings	247
B	Summary of Relevant Reports and Documents	250
C	Members of the Committee	272

REALIZING THE POTENTIAL OF C4I

Executive Summary

THE CHARTER

Section 262 of the Defense Authorization Act for Fiscal Year 1996 directed the Secretary of Defense to request that the National Research Council conduct a review of current and planned service and defense-wide programs for command, control, communications, computers, and intelligence (C4I) with a special focus on cross-service and inter-service issues. (For purposes of this report, C4I systems include systems designed to support a commander's exercise of command and control across the range of military operations and to generate information and knowledge about an adversary and friendly forces.)

THE CONTEXT

Although the Cold War is over, new regional threats to U.S. interests are increasingly likely. The U.S. military, in its traditional role as an instrument of national power, will be required to deal with a more varied set of military tasks and missions, helping to both establish and maintain regional peace and stability and also coping with less traditional tasks such as humanitarian relief and disaster recovery. Budget pressures have already resulted in a significantly reduced force structure and withdrawal of U.S. military presence from many overseas locations. Joint operations are now the norm, and in many cases, U.S. military operations are combined with those of allied and coalition forces. Forces responding to contingencies are likely to be employed "come as they are," with only minimal time for preparation and deployment before entering the operational phase of a contingency.

The military that must play these roles has many different C4I systems, both old and new. Older systems often were built for single-purpose, stand-alone applications, and often rely heavily on military-specific technology. In contrast, current systems are increasingly being built to meet explicit requirements for interoperability and flexibility, and the Department of Defense (DOD)[1] has been increasingly capitalizing on commercial information technologies for C4I systems. DOD's focus on using C4I as a way to empower the forces is an approach made easier by the fact that more and more military personnel are familiar with information technology.

THE POTENTIAL

To make a smaller military force more effective, DOD is planning to rely more than ever before on the use of high-technology C4I systems to leverage its military assets. DOD's vision of the future—Joint Vision 2010—is one of information superiority.[2] In this vision, combat planning and execution are much faster, and smaller forces are much more autonomous and lethal. Integrated C4I systems, which exchange data and work together, help military forces to prevail against adversaries by operating in a rapid, coherent, and coordinated fashion never previously achieved. Commanders at all levels can control their forces and apply their weapons with a high degree of precision, certainty of location, and awareness of the environment and of enemy actions and intentions. Responsive and reliable information technology provides timely intelligence, greater situational awareness, and a single integrated operational picture of the battlefield.

THE CHALLENGES

Joint Vision 2010 is compelling, but unrealized. The evidence to support it comes from a host of sources, including analysis, simulation, ex-

---

[1] According to Department of Defense Directive 5100.1, promulgated September 25, 1987, the Department of Defense is composed of "the Office of the Secretary of Defense (OSD), the Military Departments and the Military Services within those Departments, the Joint Chiefs of Staff (JCS) and the Joint Staff, the Unified and Specified Combatant Commands, the Defense Agencies and DOD Field Activities, and such other offices, agencies, activities and commands as may be established or designated by law, or by the President or the Secretary of Defense." This report adopts this convention, and the use of the term "DOD" without other qualification refers to all of the constituent elements described in this directive.

[2] Chairman of the Joint Chiefs of Staff. 1996. *Joint Vision 2010*, Joint Chiefs of Staff, Washington, D.C.

periments, and experience from the private sector and DOD. These sources suggest that information technology, and by extension C4I systems, can enable entirely new modes of military operation with much greater military effectiveness, just as they have radically changed how many businesses operate. These possible new modes include greater freedom of action for small, decentralized forces and the massing of firepower rather than massing of forces.

However, the vision is as yet unrealized, because it is not yet known how to exploit information technology across the full spectrum of military operations. Realizing the benefits of new C4I technologies may well require trade-offs between the C4I systems acquisition and other force investments, as well as requiring major changes in doctrine. DOD's goal must be improved military effectiveness, not simply improved capabilities. In addition to sound military judgment, careful analysis of results from well-instrumented simulations and exercises is needed to evaluate the impact of information technology, and to drive budget trade-offs between C4I and other systems.

A related issue is that new C4I systems are based on rapidly advancing computing and communications technology, driven primarily by the commercial sector. Rapid advances usually mean rapid obsolescence, so technology exploitation must be a continuous process if superiority is to be maintained vis-à-vis potential adversaries who have access to the same underlying information technologies. Both military doctrine for C4I and the budget mix of C4I versus weapons must be periodically reevaluated.

DOD policy and strategy clearly recognize the potential value of C4I technology in enhancing military effectiveness, and a number of activities and initiatives under way, both within the services and, to a lesser extent, in the joint arena seek to realize this potential. Most prominent, of course, is Joint Vision 2010.

The committee sees three major challenges to the effective exploitation of the potential offered by C4I technology—interoperability, information systems security, and DOD processes and culture involving C4I. This report is focused on these three challenges. While all three challenges are important ones for DOD to address, the committee calls attention to the security challenge (including related process and culture issues) as posing a high level of current risk. In contrast, failure to fully exploit the potential leverage of C4I represents a longer-term risk; success depends on meeting the challenges of interoperability and DOD processes and culture with respect to acquisition and effective use of C4I technologies.

DOD has recognized the importance of these challenges in various directives and initiatives. But the totality of the DOD response to these challenges is not adequate to fully exploit C4I technologies. Furthermore, it is unrealistic to expect to address these challenges "once and for all." Rather, meeting these challenges will demand continuous attention and effort over time. (In more colloquial terms, each of these areas can be regarded as a partially filled glass. The level of water in the glass represents the extent to which DOD goals for C4I have been achieved.[3] Today, the glass has both some leaks (representing matters becoming worse and failures to make progress) and a faucet putting water into the glass (representing DOD efforts to make progress). One could also argue that the glass is growing larger, representing the rapid increase in the capabilities that the technologies afford. A one-shot effort, no matter how massive, will eventually leak out. Thus, the challenge is to close up existing leaks (even as new leaks open up), and open the spigot on the faucet wider.)

THE APPROACH

The three major challenges—interoperability, information systems security, and DOD process and culture—are discussed in more detail below. For each area, a high-level goal is stated. Principles relevant to achieving that goal then follow; these principles are derived primarily from the committee's professional experiences and expertise in the civilian and military worlds set against what the committee saw and learned in its briefings and site visits. The committee's findings in each area are based on what the committee learned in the briefings it received and in the site visits it conducted, against the backdrop of these principles. Finally, specific and actionable recommendations in each area are made.

The principles and the findings and recommendations have different time horizons. The latter are tied to "today," that is, to the specific time frame in which the committee undertook this study. Five years from now, they may well no longer be timely. By contrast, the principles are intended to be more enduring, in that they frame useful questions that can be asked of DOD's efforts in C4I both today and in the future.

Because the recommendations are intended to be actionable today, the committee tried to identify specific offices that could take management action to make something happen. On the other hand, DOD—especially the Office of the Secretary of Defense and the Joint Chiefs of Staff—is engaged in an ongoing restructuring and streamlining effort. Thus, while the recommendations do identify action offices that the committee believes are appropriate, the intent is to focus more on *what* needs to be done than on the details of *who* is to do it. Finally, in the interests of space,

---

[3]Note—the water in the glass does NOT represent resources, but rather what has been achieved with those resources.

the findings and recommendations are supported in this executive summary by highly condensed versions of the argument and explanation that accompany them in the main text. Readers are urged to consult the main text for more detailed support.

C4I INTEROPERABILITY

Goal: Operational and technical interoperability commensurate with the role of C4I in support of multi-unit, joint, and combined missions.

Joint, flexible, and coherent operations are key components of DOD's vision (e.g., Joint Vision 2010); this means operational interoperability of forces and technical interoperability of C4I systems. Future U.S. military operations will inevitably involve elements from more than one service. Forces will probably be assembled with minimal time for planning and deployment, in ad hoc configurations, and for geographically far-flung missions that are highly diverse compared to those undertaken during the Cold War (and thus less predictable in advance). To enable fast and effective responses, interoperability must be built into the force structure across service and unit boundaries. Achieving adequate C4I interoperability is inherently a distributed, horizontal challenge that must be addressed in a largely vertical world. This means that there must be incentives and rewards for investments and actions across organizational boundaries.

Principles

• *The needs of the operational military commander must be the main driver of interoperability solutions and investments.* These needs exist both at the higher levels of command (e.g., the specified unified commander-in-chief or the joint task force commander) and at the tactical level where the services work together to accomplish joint missions. Interoperability is valuable not for its own sake, but only when it helps to accomplish a mission.

• *While universal interoperability is neither necessary nor achievable, a high degree of interoperability is needed to provide the flexibility required for both anticipated mission needs and unanticipated operational deployments.* What specific operations must be anticipated? Some are reasonably clear today (e.g., U.S. war plans for responding to a North Korean attack in the Korean peninsula define a specific operational context for U.S. and allied forces). Even when the theater of operations is not known, certain mission needs are likely (e.g., the need to ensure air superiority or to provide defenses against ballistic and cruise missiles). But other contingencies in which U.S. forces will be deployed will be unexpected, which places a premium on flexibility in the operational capabilities of U.S. forces—including interoperability.

- *Interoperability must be balanced against other fundamental attributes of C4I systems, including security, availability, flexibility, survivability, and performance.* Military commanders need many things from their C4I systems besides interoperability, and trade-offs among these needs are often required.
- *C4I interoperability requires a unifying framework and a body of definitive implementing guidance.* The C4I "system of systems" is large, complex, and distributed across organizational, program, and geographical boundaries. A framework and guidance are crucial because achieving C4I interoperability is largely a matter of management, design, and implementation discipline rather than of resolving technical issues. To date, the DOD has partially codified the framework and guidance in the still-incomplete architectural triad of technical, systems, and operational architectures.
- *When developing architectures, use a small team.* Good architects are critical in developing a good architecture. The role is demanding, requiring an ability to balance needs and resources, technologies, and the interests of multiple stakeholders. Good architectures usually result only when a small number of people are responsible for their content and structure. Good architectures are unlikely to emerge from a large team or from a broad consensus-based approach. These almost always involve compromises that lead to excessive complexity rather than a clear design philosophy, which in turn confuses the implementers.
- *Decompose the problem of achieving defense-wide interoperability into manageable pieces.* This principle arises from three underlying factors. First, the domain must be sufficiently bounded that progress can be made before the key players, mission requirements, or technology change significantly. An effort that is too large will simply never reach closure. Second, the problem to be addressed must not be overly complex. Third, the small teams required by the previous principle can only undertake problems of limited scope. The defense-wide network of systems, and the full spectrum of missions, are simply too large to be approached in one single effort.
- *Assess interoperability on the basis of ongoing training and testing.* Using standards makes interoperability among C4I systems easier, but does not guarantee it. Standards do not provide a complete design specification. Furthermore, given the continuing, asynchronous fielding of new systems and capabilities, interoperability is a time-perishable commodity. Only ongoing testing of a C4I system throughout its life cycle will ensure interoperability. This must include training and testing across a wide range of possible configurations that includes the other C4I systems with which it is expected to interoperate.

- *Measure progress toward interoperability goals.* Measurement and assessment—and reporting of results in a visible way—are essential to continued focus and to setting the right priorities (an instance of the general measurement principle that is articulated below under the process and culture goal). Despite laudable case-by-case efforts, there is today no method for tracking interoperability on a comprehensive or systematic basis.
- *Build a common defense-wide infrastructure to facilitate interoperability.* Where common systems and software are used, it is easier to make them interoperate. Common infrastructure is not a cure-all, however. It will not, for example, address some user or mission-specific needs.
- *Engineer for flexibility.*
  — *Use commercial off-the-shelf (COTS) products, services, and technology whenever possible.* COTS products and services improve quickly, are more sustainable, and are usually less expensive than those custom-provided to the military. When commercial products are used, the vendor often assumes much of the burden of ensuring interoperability and backward compatibility. Decisions to use COTS products and services must, however, take into account possible security risks.
  — *Use standards.* Technical standards are one way of planning for the future. Compliance with technical standards is an investment that makes future interoperability easier, though by no means certain.
  — *Base architectures and system designs on layering and clean interfaces.* Layered architectures make it possible to exploit technological progress in some parts of a system without the need for a total system redesign. Clean interfaces make it easier to interoperate with other systems that conform to those interfaces. Interfaces are an investment in the future: by providing well-defined ways to access systems and capabilities, they make it easier to compose these components in new ways in the future, or to use existing systems in new ways.
  — *Make data self-describing to permit future interoperability.* Another investment in future interoperability is to identify the meaning of data so that it can be used in future applications. Examples include recording and transmitting not only a position but also the coordinate system it is given in, or generating a time stamp for a target track to help other systems resolve multiple tracks.

Finally, because the analysis of and solutions to interoperability problems are inherently distributed throughout and across the DOD, interoperability efforts should be guided by a final principle:

- *Achieving interoperability requires responsibility and authority that crosses organizational boundaries—a requirement that implies the need for strong top-down leadership.* This crossing of boundaries is particularly important to the development and fielding of systems that support joint operations, as well as to the development of doctrine for joint operations. The DOD must search for practical ways to reward interoperability and impose sanctions for ignoring it. Sanctions are unwieldy and can be applied only at great cost and effort, and only in a few cases. Therefore, although they do have value in focusing attention on flagrant offenders, it is much better in the long term to establish a culture that rewards interoperability.

Findings

Parts of DOD are well aware of a defense-wide problem in exploiting rapidly changing information technologies, in using commercial off-the-shelf products effectively, and in security. There are in place today a DOD strategy and ongoing efforts to promote interoperability, resting on technical standards such as the Joint Technical Architecture and the use of a defense-wide common infrastructure. While much has been accomplished, the goal of a C4I system of systems with assured interoperability for the U.S. military continues to be unachieved. Progress has in many cases been slow, and past C4I studies[4] show that many documented C4I interoperability problems remain unresolved. Despite increased attention and management awareness, much more must be done before C4I interoperability is sufficient to provide adequate end-to-end support of military missions and cease being a major constraint on the execution of military operations.

Finding I-1: While the elements of DOD's current strategy for achieving interoperability are positive, they are not being fully executed. Both formulation and implementation have gaps and shortfalls.

The DOD technical interoperability strategy (adopting an architectural approach, building to standards defined by the Joint Technical Architecture, and developing a common, defense-wide "public utility" infrastructure) builds on the best practice in industry and is a very important step that promises to significantly improve interoperability over time. At the same time, this strategy is not being fully executed. There has been insufficient progress in the development and implementation of the Joint

---

[4]See, for example, General Accounting Office. 1993. *Joint Military Operations: DOD's Renewed Emphasis on Interoperability Is Important But Not Adequate*, General Accounting Office, Washington, D.C.

Systems and Joint Operational architectures, in ensuring compliance with the Joint Technical Architecture, and in building and using a common infrastructure.

Finding I-2: Even full execution of the DOD strategy for interoperability will not assure that joint mission needs for C4I will be met.

First, priorities must be set and the problem bounded in size to make it more manageable. Second, interoperability must be built in throughout the life cycle of C4I systems—in development, initial fielding, ongoing assurance, and resolving problems faced by deployed forces. Third, there must be a system to measure the interoperability of C4I systems, both for assessing progress in development and acquisition and for assessing the interoperability component of force readiness. Fourth, there must be concrete guidance on technology evolution and the role of COTS technology. Finally, neither the DOD-wide mandatory Enterprise Data Model Initiative[5] nor the voluntary collaboration approach to data interoperability embodied in the Shared Data Environment (SHADE) program is likely to be adequate.

Recommendations

Some of the interoperability challenge stems from the broader issue of the distributed, horizontal structure and organization of DOD itself, as established by Title X. The recommendations that follow do not assume any changes to this fundamental framework. While the specifics of these recommendations are directed at achieving interoperability among U.S. forces, the principles they embody do apply to interoperation with at least some coalition partners—those who are members of an existing alliance framework. However, management is clearly much more complex when several nations are involved.

Recommendation I-1: The Assistant Secretary of Defense for C3I and the Joint Chiefs of Staff should complement the DOD's current broad interoperability strategy with focused efforts in limited, operationally important domains, to include the development of Joint Operational and Joint Systems architectures for these domains.

An all-at-once development of an operational architecture covering the entire span of DOD's operational requirements is not feasible. Opera-

---

[5]Department of Defense Directive 8230.1-M, "DOD Data Administration," September 26, 1991.

tional architectures must instead be developed for particular joint missions or tasks, organized either around significant operational capabilities or around mission slices. These slices or capabilities should be operationally important, be inherently joint, involve a large enough number of systems to warrant the effort, and be ones where significant foundational work has already been done. The focused activities would complement the defense-wide standards and common infrastructure initiatives that provide a necessary foundation for mission-specific capabilities.

Recommendation I-2: The Secretary of Defense should establish a joint C4I integration and interoperability activity to address integration and interoperability throughout the entire life cycle of C4I systems.

Current DOD activities for promoting C4I interoperability should be augmented in three areas: cross-service testing starting early in the development process, ongoing interoperability assurance in operational contexts, and interoperability support for deployed forces. The joint C4I integration and interoperability activity would do this work, taking development, testing, and training roles in peacetime and providing support during exercises and operational deployments.

Recommendation I-3: The Secretary of Defense, the Assistant Secretary of Defense for C3I, and the Chairman of the Joint Chiefs of Staff should establish processes to assess C4I interoperability on a regular basis.

Recommendation I-3.1: The Assistant Secretary of Defense for C3I and the Joint Chiefs of Staff should develop a set of "interoperability scorecards" as a basis for management, covering the spectrum from compliance with standards to successful end-to-end mission support.

Three scorecards are proposed—technical, systems, and operational—corresponding to the elements of the architectural triad. A technical compliance scorecard assesses how well systems comply with defined interoperability standards and guidance. A systems interoperability scorecard measures actual interoperability between C4I systems. An operational interoperability scorecard measures the ability of a set of systems to satisfy the information flows needed for a particular mission.

Recommendation I-3.2: The Chairman of the Joint Chiefs of Staff should establish a process to incorporate C4I interoperability into readiness reporting.

Although individual combat units can report their combat readiness, they often cannot assess their interoperability readiness. The readiness of

C4I systems must be assessed at higher echelons of command, particularly those with a joint perspective. Today no formal combat readiness reporting system exists at these levels. The system that the Chairman of the Joint Chiefs of Staff develops must focus on assessing the ability of forces to conduct end-to-end missions, based on a realistic set of scenarios for how units are to be employed. It may be appropriate to focus assessment efforts in the same mission slices as those in which the activities proposed in Recommendation I-1 are conducted.

Recommendation I-4: The services and agencies should designate an activity within the program offices for C4I systems (and weapons systems with embedded C4I) to be explicitly responsible for resolution of architectural and system-level issues that determine interoperability.

An "interoperability cell" or equivalent in C4I program offices would provide a central point of focus for interoperability issues, with an outward-looking cross-service perspective. Such an activity would provide a "bottom-up" approach to interoperability to complement "top-down" architectural and common infrastructure efforts. The cell would be responsible for revising architecture as needed to accommodate changes in doctrine, tactics, techniques, procedures, and equipment; engaging the stakeholders in a particular C4I system in making that system interoperable; and negotiating interoperability issues with those responsible for "neighboring" systems.

C4I SYSTEMS SECURITY

Goal: C4I systems that remain operationally secure and available for U.S. forces in the face of attacks by adversaries.

The more military leverage that C4I systems provide for U.S. forces, the larger the incentives are for an opponent to attack those systems. Indeed, it makes little sense for an opponent to challenge the United States symmetrically, i.e., force on force. More likely avenues of challenge are asymmetric ones that exploit potential U.S. vulnerabilities. Attacking U.S. C4I systems—whether directly or indirectly (e.g., through the U.S. civilian information infrastructure on which DOD C4I systems often depend)—is only one of many possible asymmetric attacks, but one for which the United States must be adequately prepared. Because the DOD understands the challenges of physical security for C4I systems very well, this report focuses on cyber-security.

Principles

- *A culture of information security is required throughout the organization.* The culture of any organization determines how seriously its members take their security responsibilities. For information security, policies and practices are at least as important as technical mechanisms. Policies specify the formal structures, ensure responsibility and accountability, establish procedures for using technical means of protection and assigning access privileges, create sanctions for breaches of security at any level, and require training in the relevant practices and use of security technologies. Furthermore, senior leadership must take the lead to promote information assurance as an important cultural value. Top-level commitment is not sufficient to ensure good security practices. Without it, however, organizations will not focus on security but will expend their energy on other things that seem more directly related to their core missions.
- *Cyber-attack is easier than cyber-defense.* The reason is that effective defense must be successful against all attacks whereas an attacker need succeed only once. Cyber-attack is easier, faster, and cheaper than cyber-defense. Paradoxically, cyber-attack appears to be more highly rewarded in U.S. military culture. Consequently, experts in cyber-attack are more numerous than those in cyber-defense. Today, the need for cyber-defenders far outstrips the supply, and defenders must be allocated wisely, encouraged in their efforts, and increased in their numbers.
- *Cyber-attackers attack the weakest points in a defense*, and every system has weak points. ("An army is like water: it avoids obstacles and flows through low places.") Thus, the security of a system—any system—can never be guaranteed. Any system is always compromised to some extent, and a basic design goal of any system must be that it can continue to operate appropriately in the presence of a penetration. Vulnerabilities include fraudulent identification and authorization, abuse of access privileges, compromises in the integrity of data or programs, and artificially induced disruptions or delays of service.

Implementation of good system security depends on several principles:

- *Defend in depth.* Defense in depth is a sound countermeasure against security failures at a single point and also against security failures that share a common mode. Furthermore, an attacker that faces multiple defenses must have the expertise to overcome all of them (rather than just one) and must also expend the time required to overcome all of them.
- *Ensure graceful degradation of compromised systems.* Prudence requires C4I developers and operators to assume some non-zero probability that any system will be successfully attacked, that some DOD systems have

been successfully attacked, and that some C4I systems are compromised at any given moment. Nevertheless, most of the C4I systems connected to compromised components (and organizations that rely on these systems) should be able to function effectively despite local security failures.

- *Manage the tension between security and other desirable C4I attributes*, including user convenience, interoperability, and standardization. This tension is unavoidable. The desire for any of these attributes should not be used as an excuse for not working on security, or vice versa. From an acquisition standpoint, security is currently too often regarded as an afterthought in the design and implementation of C4I systems.
- *Do what is possible, not what is perfect.* Insistence on "perfect" security solutions for C4I systems means that as a practical matter, C4I systems will be deployed without much security functionality. By contrast, a pragmatic approach that makes significant use of commercial information security products and provides moderate protection is much better than nothing. In this respect information security is very different from communications security, because information systems are much more complex.
- *Recognize the inherent weaknesses in passive defense.* Because passive defense techniques are used to provide security, an unsuccessful attack on a C4I system usually does not result in a penalty for the attacker. Thus, a persistent attacker willing to expend the time to find weaknesses in system security will eventually be successful. Cyber-defenders of C4I systems must anticipate facing persistent attackers.

Findings

Finding S-1: Protection of DOD's information and information systems is a pressing national security issue.

DOD is in an increasingly compromised position. The rate at which information systems are being relied on outstrips the rate at which they are being protected. Also, the time needed to develop and deploy effective defenses in cyberspace is much longer than the time required to develop and mount an attack. The result is vulnerability: a gap between exposure and defense on the one hand and attack on the other. This gap is growing wider over time, and it leaves DOD a likely target for disruption or pin-down via information attack.

Finding S-2: The DOD response to the information systems security challenge has been inadequate.

In the last few years, a number of reports, incidents, and exercises have documented significant security vulnerabilities in DOD C4I systems.

Despite such evidence, the committee's site visits revealed that DOD's words regarding the importance of information systems security have not been matched by comparable action. Troops in the field did not appear to take the protection of their C4I systems nearly as seriously as they do other aspects of defense. Furthermore, in many cases, DOD is prohibited by law and by national policy from taking retaliatory action against a cyber-attacker that might deter future cyber-attacks. On the technology side, information systems security has been hampered by a failure to recognize fully that C4I systems are today heavily dependent on commercial components that often do not provide high levels of security. Furthermore, the C4I security practices that the committee observed in many of its site visits were far inferior to the standard set by the best DOD and private-sector practices for information systems security. Given the importance of DOD C4I systems to the national security and the sensitivity of the information handled in those systems, the committee would have expected DOD C4I security practices, in general, to reach a higher standard than was found.

Recommendations

The committee believes that operational dimensions of information systems security have received far less attention and focus than the subject deserves in light of a growing U.S. military dependence on information dominance as a pillar of its warfighting capabilities. Furthermore, the committee believes that *it is urgent that DOD greatly improve the execution of its information systems security responsibilities.*

One critical aspect of improving information systems security is changing the DOD culture, especially within the uniformed military, to place a high value on it. With a culture that values the taking of the offensive in military operations, the military may well have difficulty in realizing that defending against information attack is more critical and more difficult than conducting an information attack against an adversary. Senior DOD leadership must therefore take the lead to promote information systems security as an important cultural value for DOD. The committee was encouraged by conversations with several senior defense officials, both civilian and military, who appeared to take information systems security quite seriously. Nevertheless, these officials will have a limited tenure, and the need for high-level attention is a continuing one.

A second obstacle to an information systems security culture is that from an operational perspective good security often conflicts with getting things done. And because good information systems security results in nothing (bad) happening, it is easy to see how the can-do culture of DOD might tend to devalue it.

Recommendation S-1: The Secretary of Defense, through the Assistant Secretary of Defense for C3I and the Chairman of the Joint Chiefs of Staff, should designate an organization responsible for providing direct operational support for cyber-defense to commanders.

Defensive information operations require specialized expertise that may take years to develop. This means that in the short run it is unrealistic to expect operational units to develop their own organic capabilities in this area. An organization that supports all commanders would bring specialized defensive expertise to bear in both exercises and real military operations. Close coupling between operators and the information systems security arena is a necessary precondition for achieving adequate security in fielded systems.

Recommendation S-2: The Secretary of Defense should ensure that adequate information system security tools are available to all DOD civilian and military personnel, direct that all personnel be properly trained in the use of these tools, and then hold all personnel accountable for their information system security practices.

Accountability for upholding the values of an organization is an essential element of promulgating a culture. Once senior leaders have articulated a department-wide policy for information assurance and provided personnel with appropriate tools, it is necessary to develop well-defined structures with clear lines of responsibility. Accountability depends on the availability of adequate tools that make good security possible with reasonable effort; ongoing education and training in security practices; incentives, rewards, and opportunities for professional advancement for promoting compliance with good security practices; continuous measurement of security; and sanctions for violations of good information assurance practice that are applied uniformly and consistently to all violators, regardless of rank.

Recommendation S-3: The Secretary of Defense, through the Assistant Secretary of Defense for C3I, the Chairman of the Joint Chiefs of Staff, and the CINCs,[6] should support and fund a program to conduct frequent, unannounced penetration testing of deployed C4I systems.

---

[6]CINC, an acronym for "commander-in-chief," refers to the commander of a specified or unified combatant command. The term "CINCs" refers to the commanders of the combatant commands. The combatant commands include the U.S. European Command, U.S. Pacific Command, U.S. Atlantic Command, U.S. Southern Command, U.S. Central Command, U.S. Space Command, U.S. Special Operations Command, U.S. Transportation Command, and U.S. Strategic Command.

Because all systems have technical and operational vulnerabilities (and develop new ones as they evolve), a continuing search for those weaknesses is essential. Only independent and unscheduled "red team" probes provide reliable information on actual weaknesses. This information can be used to enforce accountability for good security practices or to focus attention on necessary technical or procedural fixes, depending on the source of the weakness. Note the critical focus on C4I systems that are operating in a "full-up" mode, rather than on individual C4I components.

Recommendation S-4: The Assistant Secretary of Defense for C3I should mandate the immediate department-wide use of currently available network and configuration management tools and strong authentication mechanisms.

DOD-wide use of proper configuration management tools and strong (non-password) authentication mechanisms would be an important step toward upgrading the security of DOD C4I systems to the level of best practices in the private sector. Network management tools can continuously monitor the operational configuration of a network and all of its component machines, alerting the administrator when variances from known (and safe) configurations are detected. Strong authentication mechanisms nearly eliminate the vulnerabilities of passwords for authentication. Furthermore, they can also be used to authenticate all computer-to-computer communication; thus all communications carried in the network can be authenticated rather than just those originating from outside a security perimeter.

Recommendation S-5: The Under Secretary of Defense for Acquisition and Technology and the Assistant Secretary of Defense for C3I should direct the appropriate defense agencies to develop new tools for information security.

Aligning DOD information security practice with the best practices found in industry today would be a major step forward in the DOD information security posture, but it will not be sufficient. Given the stakes of national security, DOD must go further. Going further will require research and development in many areas, including configuration control and systematic code verification; fine-grained authorization for resource usage; tools for adaptive or active defense; accurate and rapid location of attackers in cyberspace; secure composition of secure systems and components to support ad hoc (e.g., coalition) activities; better ways to configure and manage security features; generation of useful security specifications from programs; more robust and secure architectures for networking (e.g., requiring trackable, certified authentication on each packet, along with a network fabric that denies transit to un-authenticatable packets); and automatic determination of classification from content.

Recommendation S-6: The Chairman of the Joint Chiefs of Staff and the service Secretaries should direct that a significant portion of all tests and exercises involving DOD C4I systems be conducted under the assumption that they are connected to a compromised network.

Prudent operation of C4I systems requires C4I developers and users to assume some non-zero probability that any system will be successfully attacked, that some DOD systems have been successfully attacked, and that some C4I systems are compromised at any given moment. (A "compromised" system or network is one that an adversary has penetrated or disrupted in some way, so that it is to some extent no longer capable of serving all of the functions that it could serve when it was not compromised.) However, despite this assumption, most of the C4I systems connected to the compromised components should be able to function effectively despite local security failures. Exercises conducted under this pessimistic assumption allow the U.S. military to be trained in how to use its C4I systems and networks even if they have been compromised, as well as for the possibility that they will be largely unavailable for use at all.

Recommendation S-7: The Secretary of Defense should take the lead in explaining the severe consequences for U.S. military capabilities that arise from a purely passive defense of its C4I infrastructure and in exploring policy options to respond to these challenges.

The notion of cyber-retaliation raises many legal and policy issues, such as differences between appropriate responses in wartime and peacetime, how to respond to domestic and foreign attackers (and attackers of uncertain origin), and the role of law enforcement authorities vis-à-vis the role of DOD. As a first step, DOD should review the legal limits on its ability to defend itself and its C4I infrastructure against information attack. After such a review, DOD should take the lead in advocating changes in national policy (including legislation, if necessary) that amend the current "rules of engagement" specifying the circumstances under which force is an appropriate response to a cyber-attack against its C4I infrastructure. The committee was not constituted to address the larger questions of national policy, e.g., whether other national goals do or do not outweigh the narrower national security interest in protecting the U.S.

military information infrastructure. It is explicitly silent on the question of whether DOD should be given the authority (even if constrained and limited to specific types and circumstances) to allow it to retaliate against attackers of its C4I infrastructure. But the committee does believe that DOD should take the lead in explaining the severe consequences for its military capabilities that arise from a purely passive defense.

DOD PROCESS AND CULTURE

Goal: A DOD culture and management system that fully reflects the importance of C4I in future military operations and the pace at which the underlying technologies evolve.

While both C4I interoperability and C4I security have technical and non-technical elements, DOD process, culture, and military doctrine are not issues of technology per se. Rather, they are issues of management and how to exploit the leverage afforded by technology as fully as possible. Realizing the full potential offered by Joint Vision 2010 will require significant doctrinal innovations that combine technology with new operational concepts. At the same time, just as many private-sector attempts at reengineering fail, new doctrines, new modes of operation, or new tactics may look promising but be unsound in fact. Thus, continuing exploration and experimentation are needed to validate major changes in these areas. In addition, the pace of progress in the underlying information technologies means that internal DOD processes to deal with the acquisition of C4I systems—as well as the trade-offs in emphasis among C4I, weapons systems, and personnel—will have to be changed radically if the DOD is to fully exploit advances in information technology. Joint Vision 2010 provides a top-level vision of what C4I technology can do for military operations, but the road from vision to realization is quite rocky, and progress has so far been too slow. DOD is changing, but it is not changing fast enough to fully exploit the opportunity for information superiority.

Principles

- *Cultural change requires a clear vision of what is to be, together with processes that refine and communicate the vision.* A clear vision is the essential starting point for changing organizational culture to take advantage of information technology.
- *The senior leadership of the organization must be persistently, visibly, and deeply committed to driving cultural change.*
- *The organization must be willing and able to reengineer key processes* in order to take maximum advantage of technology.

- *The organization must be willing and able to reallocate resources commensurate with its vision*, because the introduction of new technology is often expensive in the short term (both for procurement and training).
- *The organization must systematically measure progress and change in its organizational processes, results, and performance of key people.* Measurements are needed so that the organization can understand what remains to be done in achieving its goal.

Special attention must be paid to the DOD acquisition system and the human resource base. If DOD is to effectively exploit rapidly evolving information technologies, the acquisition system for C4I systems must take due account of several principles:

- *Accept the "80% solution."* Because users are often unable to specify exactly what they want until they see it, implementing an 80% solution provides a useful point of departure from which users can articulate their needs more precisely. Furthermore, an 80% solution provides immediately useful functionality, as well as benefits in the form of cost reduction and time to delivery.
- *Accept and manage risk in oversight processes.* Because information technology changes so rapidly, investments in C4I systems are inherently risky. They enable new ways of conducting military operations that may be at odds with established doctrine, and if not managed properly they run the risk of being obsolete before they are available for use. Decision makers will never have anything approximating perfect knowledge of how a C4I system will be used, and so risks must be accepted as part of the decision-making process.
- *Test C4I systems cooperatively, collaboratively, flexibly, and continuously.*
- *Exploit experimental programs.* Only through experiments can new C4I-enabled modes of military operations be discovered and explored, and their implications for C4I use understood.
- *Seek budget flexibility.* Especially in the context of a 5-year defense budget plan, funding should be promptly available to take advantage of unanticipated C4I applications.

The human resource base of the DOD is also critical to the effective exploitation of information technology. The following principles are important:

- *Technology specialists and combat operators must be knowledgeable about both operations and technology.* Combat operators should be deeply knowledgeable about the present and projected capabilities offered by

C4I systems, and C4I specialists should be deeply knowledgeable about combat operations. The full potential of C4I can be achieved only by exploiting the synergy between operations and technology. Such reciprocal knowledge about one another's domains will undoubtedly require cross-training.

- *Career paths in DOD must provide competitive reward, professional challenge, development, and recognition.* DOD must compete with the private sector for information technology expertise, and while it cannot offer compensation packages equal to those found in the private sector, it must go out of its way to reduce the differential.

Findings

Finding P-1: DOD processes dealing with the acquisition of C4I systems have not been adequately restructured to account for the rapid pace of development in the commercial information technologies on which such systems will inevitably build.

The current acquisition system is particularly ill-suited to C4I systems. First, program management and oversight processes are heavily weighted toward metrics associated with historical acquisition methods associated with weapons systems in which the underlying technologies change much more slowly. Second, DOD no longer enjoys the leverage it once had in developing and applying information technology. Thus, C4I systems—unlike most weapons systems—increasingly rely on commercial technologies. Third, the current acquisition process assumes that a service can identify a specific system or program to address specific and articulated military needs. Such an assumption may be reasonable for weapons systems, but it is inadequate for C4I systems for two reasons. One reason is that C4I systems, and especially infrastructure such as networks, are often more valuable in enhancing the capability of several weapons systems than in meeting specific needs. A second reason is that C4I users more often come to understand their requirements by experimenting with prototypes than by deep intellectual analysis conducted on paper. Finally, acquisition personnel have not been well trained to manage C4I acquisitions or socialized into an information technology culture.

Finding P-2: In many instances, operational processes do not appear to have been reengineered to take full advantage of the capabilities that C4I technology can provide.

Commercial experience strongly suggests that the maximum benefits of information technology come not from automating existing business processes, but rather from developing new processes that take full advantage of the new technologies. Such reengineering is quite difficult both for the private sector and for the DOD. When successful, however, reengineering gives enormous leverage. The competitive arena for the military is not as well defined as that for private-sector enterprises, but reengineered, technology-exploiting operational processes should yield major competitive advantage in the military, driving revisions of doctrine, smaller logistics footprints, enhanced agility, and a redefinition of the skill set required in the fighting forces. In its site visits and briefings, the committee saw a wide range of organizational responses to C4I technology. In some cases, new modes of combat operations were being explored and potential points of high leverage found. However, in most cases that the committee observed, C4I technology was being used to speed up existing processes. Some benefits were apparent from these latter efforts, but incremental application of technology in this way seldom results in large (order-of-magnitude) benefits.

Finding P-3: The military services have not accorded to information technology and C4I professionals stature comparable to their increasing importance for battlefield operations.

Well-trained C4I professionals are essential to the successful operation of modern military weapons such as jet fighters, warships, and sophisticated ground-based weapons. However, DOD is not succeeding in creating either the environment or the incentives to attract and retain such human resources. One problem is that DOD has not yet found a way to integrate its C4I personnel into combat line elements and to make them fully conversant with military doctrine, strategy, and tactics. Rather, they are often regarded as implementers of high-level strategy decisions that are made without their input, and the status and prestige of C4I specialists are not comparable to those of personnel in traditional combat arms specialties. Furthermore, the DOD culture tends to discourage attracting and retaining the necessary engineering, system integration, and applications talent for implementing and sustaining high-technology C4I systems. The private sector can offer greater monetary rewards, personal recognition, and opportunity for advancement, and thus beckons to every engineer, technician, and system specialist in the military—enlisted or officer.

Finding P-4: The DOD process for coupling end-user operational needs to C4I systems is inadequate.

The general principle that operational needs should drive the acquisition system is well established within DOD. But under the traditional acquisition system, warfighter input (from the perspectives of the CINCs) enters the acquisition process only at the start of a new program. Thus, input from the end users—the field commanders—cannot easily be accommodated, because it is generally infeasible to specify requirements for C4I systems in a form that they can be handed "over the transom." Furthermore, warfighter input (especially that from a joint perspective) can be diluted when individual services are responsible for setting system requirements.

Finding P-5: Achieving C4I interoperability is more a matter of organizational commitment and management (including allocation of resources, attention to detail, and continuing diligence) than one of technology.

Many parties alleged to the committee that higher degrees of C4I interoperability would require additional funding. While this is undoubtedly sometimes true, major cost savings are possible in the development of a system by reusing existing work (whether manifested as preexisting military technology or COTS technology). Most importantly, total life-cycle costs may well be less when the need to hedge against unanticipated needs for interoperability is factored in, because retrofitting systems for interoperability results in working such problems case by case, providing expensive curative rather than inexpensive preventive medicine. Finally, interoperability can make it easier to use existing resources efficiently. The committee believes that senior DOD leaders, both civilian and military, take interoperability challenges quite seriously. But DOD is not establishing a culture supportive of C4I interoperability that will outlive today's senior leaders. Without such a culture, DOD efforts to promote and enforce interoperability will be fragile.

Recommendations

Recommendation P-1: The Secretary of Defense, working with the service Secretaries and the Chairman of the Joint Chiefs of Staff, should establish in each of the services a specialization in combat information operations, provide better professional career paths for C4I specialists, and emphasize the importance of information technology in the professional military education of DOD leadership.

Today, the treatment of the technical force in DOD relegates C4I specialists to the second-class status of support, rather than line functions. If it is true that information is critical to modern warfare, and that information dominance can provide the operational military advantages of large forces without their costs, then C4I specialists must be better aligned with those in the mainstream operational community. Furthermore, senior commanders must have a good understanding of how best to exploit C4I to enhance military operations. Information system employment must become a first-line combat function, just like employment of combat forces and weapons. C4I specialists must be trained in the doctrine, strategy, tactics, and combat employment of military forces, and be fully integrated into combat units and operational planning elements of the military forces. DOD should also provide increased opportunities for promotion and recognition, as well as higher pay scales, for C4I specialists.

Recommendation P-2: The Under Secretary of Defense for Acquisition and Technology should train its civilian and military personnel who participate in the acquisition of C4I systems to understand the difference between C4I systems and weapons systems.

Program managers must understand the intrinsic differences between C4I and weapons technologies, and they must be able to argue the significance of those differences in front of acquisition boards and oversight councils that are more accustomed to dealing with weapons systems. Today, conservative "by the book" approaches that are better suited to long-lived weapons systems are regularly applied to C4I systems, even though existing acquisition rules allow considerable flexibility in the management of a C4I program.

Recommendation P-3: In order to explore and develop ("incubate") new ideas for the use of information technology to support military needs, the Secretary of Defense should establish an Institute for Military Information Technology either as a free-standing unit or by expanding the charter of an existing institution.

All levels of the DOD/service hierarchy contain individuals with good insights about existing problems, ideas about how to fix those problems, and innovative concepts about how C4I technology could be used to improve military effectiveness. But because of the traditional military command structure, those at lower levels of the hierarchy face considerable risk if they challenge the conventional wisdom. The purpose of the proposed institute would be to facilitate intellectual risk taking by bringing together for extended periods of time combat operators, military information technologists, and civilian information technology experts from academia and industry in an environment where innovative ideas for using information technology to support military needs could be explored relatively freely.

Recommendation P-4: The Assistant Secretary of Defense for C3I and the Under Secretary of Defense for Acquisition and Technology, working with the service Secretaries and the Chairman of the Joint Chiefs of Staff, should direct that as a general rule, every individual C4I acquisition should (a) use evolutionary acquisition; (b) articulate requirements as functional statements rather than technical specifications; and (c) develop operational requirements through a process that includes input from all the services and the CINCs.

Over the time scale of a typical military C4I program, the applicable technology underlying the program, as well as operational requirements for its use, the doctrine that governs its operation, and the world and local environments in which it must operate, can be expected to change dramatically. For these reasons, the initial requirement should be for an "80% solution" to the functional requirement. This will encourage the use of commercial technology and dramatically reduce the cycle time for developing new C4I systems. Furthermore, the use of functional requirements is a way to avoid overspecifications of design that limit the ability of a supplier to find better or more cost-effective ways of implementing the system. And, if all U.S. C4I systems are to be regarded as being for use in joint operations, the requirements definition process for C4I systems should be under the control of a group that represents the interests of all stakeholders. If all the services and CINCs participate in formulating requirements, not just in reviewing them, it is more likely that the system will satisfy needs for joint operation.

Recommendation P-5: The Secretary of Defense should seek, and the Congress should support, an appropriate level of budgetary flexibility to exploit unanticipated advances in C4I technology that have a high payoff potential.

As new commercial information technologies and applications emerge that can significantly improve military capabilities, management and budgeting must make it possible to exploit them. High-value C4I applications may emerge quickly (e.g., as the result of experiments or demonstrations such as the Joint Warrior Interoperability Demonstration) or on a track other than the normal acquisition track (e.g., as the result of an advanced concept technology development (ACTD)). Proper follow-on requires a process for inserting such applications into the appropriate phase of the acquisition process. Since service budgets do not include extra funds for such circumstances and reprogramming funds is a difficult task (implying that an otherwise funded program must be short-changed), an "offline" funding mechanism is required to cover unanticipated needs. Finally, even if an ACTD does not enter the mainstream acquisition process, funding streams are needed to ensure that leave-behinds from ACTDs are compatible with the other systems where they are deployed, and are maintainable and supportable.

Recommendation P-6: DOD should put into place the foundation for a regular rebalancing of its resource allocations for C4I.

Recommendation P-6.1: The Under Secretary of Defense (Comptroller) should explicitly account for C4I spending as a whole in DOD's budget process.

Because the technologies underlying C4I change so rapidly, a DOD commitment to U.S. information superiority on the battlefield of the future must be accompanied by a continuing examination of the resources allocated to C4I, especially relative to other important categories of spending such as readiness, weapons, and force structure. Because C4I is not an explicit budget category within the annual DOD budgeting process, the services for the most part determine their own C4I priorities and how those weigh against their needs for force structure and weapons procurement. Without knowing what is being spent on C4I in any given year by all the services, it is obviously difficult to make informed defense-wide overall trade-offs.

Recommendation P-6.2: The Joint Chiefs of Staff should develop and use measures of military effectiveness that can be used to assess the contribution of C4I to military effectiveness.

Spending more on C4I would necessarily mean spending less on other modernization, readiness, and force structure. DOD therefore needs to be reasonably confident that the gain attributable to C4I outweighs the loss in other areas if it moves in this direction. Quantitative measures of military effectiveness will thus be necessary to support a continuing process of rebalancing investment among C4I, weapons, and force structure (and among C4I systems themselves).

Recommendation P-7: The Secretary of Defense, the Chairman of the Joints Chiefs of Staff, the CINCs, and the service Secretaries should sustain and expand their efforts to carry out experimentation to discover new concepts for conducting information-enabled military operations.

Experimentation within the DOD context is analogous to business process reengineering in the private sector. Both seek radically new ways of doing things that create value and advance the ability of the organization to conduct military operations or to make money. Significant experimentation is under way within the DOD today. Nevertheless, it is all too easy to fall back to "business as usual" when faced with budget pressures. Experiments are undeniably expensive, and failure is to be expected from time to time. Well-meaning critics who focus on the cost and possible failure of particular individual experiments may do more damage than good in the long run. Fortunately, such criticism is rare today, but in the face of budget pressures to cut back on experimentation, the Secretary of Defense, the Joint Chiefs of Staff, and the service Secretaries will have to strongly uphold the value of investing in the future.

Recommendation P-8: DOD should develop and implement a set of management metrics that are coupled to key elements of C4I system effectiveness.

Achieving large-scale cultural change in an organization requires commensurate change in management metrics. Metrics are a major motivator of human behavior and have been demonstrated to be an essential element of making improvements: they are the base for driving continuous progress. Management metrics measure the characteristics or performance of an organization and are used by senior management to assess the effectiveness of the organization and its leadership. To assess and drive the cultural change needed to fully exploit C4I in warfighting, metrics are needed for such key areas as interoperability, security, and overall rate of implementation, as well as such associated elements as training, skill, and resource levels. These metrics should be as quantitative as possible, though in some cases judgment-based ratings will have to be used. The metrics should be applied both to units and to commanders at higher echelons in a manner consistent with their responsibilities.

CONCLUSION

Advanced C4I systems and technology offer the potential for enabling radically more effective military forces. But if this potential is to be realized, DOD will need to fix existing vulnerabilities in information systems security as well as to address challenges posed by C4I interoperability and to embrace and accommodate an information-age culture. Only through sustained action in these areas will DOD's needs for capable C4I systems be met in the coming decades.

1

Introduction

At the brink of a new century, the U.S. military is grappling with its role, its requirements, and its operational imperative as an instrument of national power. Military responsibilities span a wide range, from peacetime engagement to shape the international environment, maintain alliances, and ensure access; to stability and support operations including humanitarian assistance, disaster relief, counterterrorism, and peacekeeping; to a capability for prosecution of conflict from small-scale contingencies to major theater war. Primarily a deterrent force during the Cold War, today the U.S. military is seen more as an integral element of U.S. national power that is committed around the world on an ongoing basis. At the same time, its forces are smaller and stationed mainly in the continental United States, and the military budget will likely continue to be constrained. The resulting leaner force structure will need the versatility to project power flexibly, rapidly, and from a distance, in combination with allies and coalition forces.

Information creation, communication, analysis, and exploitation have always played a key role in military strategy and operations. But the recent and continuing rapid progress in information and communications technologies dramatically enhances the strategic role of information, positioning effective exploitation of these technology advances as a critical success factor in military affairs. These technology advances are drivers and enablers for the "nervous system" of the military—its command, control, communications, computers, and intelligence (C4I) systems—to more effectively use the "muscle" side of the military, namely the weapons and platforms and troops. The growing importance of C4I systems reflects an information technology-driven transformation of strategy and operations similar to what is occurring across almost every segment of society. Information superiority, indispensable to dominance in the full range of military operations, is central to Joint Vision 2010,[1] the conceptual template guiding Department of Defense (DOD)[2] efforts to leverage technological opportunities and structure innovations by military personnel to achieve new levels of effectiveness in joint military operations. As this report discusses in detail, in realizing this vision for C4I the U.S. military faces a fundamental set of technical and management challenges.

1.1 WHAT IS C4I?

The acronym C4I stands for "command, control, communications, computers, and intelligence" (see Box 1.1 for DOD definitions of each of these terms). Command and control is about decision making, the exercise of direction by a properly designated commander over assigned and attached forces in the accomplishment of a mission, and is supported by information technology (the computers and communications part of C4I). The United States is aggressively exploiting these technologies in order to achieve information superiority, with the objective of achieving better and faster decisions,[3] and continually projecting, albeit with uncertainties, future desired states and directing actions to bring about those future states. (Box 1.2 describes some major C4I systems; Box 1.3 describes elements of the defense information infrastructure.)

One important capability that C4I systems provide commanders is situational awareness—information about the location and status of enemy and friendly forces. A necessary component of achieving superiority in decision making, it does not alone guarantee superior decision making. Commanders must take relevant knowledge and combine it with their

---

[1]Chairman of the Joint Chiefs. 1996. *Joint Vision 2010*, Joint Chiefs of Staff, Washington, D.C.

[2]According to Department of Defense Directive 5100.1, promulgated September 25, 1987, the Department of Defense is composed of "the Office of the Secretary of Defense (OSD), the Military Departments and the Military Services within those Departments, the Joint Chiefs of Staff (JCS) and the Joint Staff, the Unified and Specified Combatant Commands, the Defense Agencies and DOD Field Activities, and such other offices, agencies, activities and commands as may be established or designated by law, or by the President or the Secretary of Defense." This report adopts this convention, and the use of the term "DOD" without other qualification refers to all of the constituent elements described in this directive.

[3]Such decisions can range from those at the theater level (e.g., deciding which forces should be deployed in what locations) to the tactical level (e.g., deciding which specific weapons should be allocated against which targets).

BOX 1.1 DOD Definitions of Terms: Command, Control, Communications, Computers, and Intelligence (C4I)

Command and control (C2)—The exercise of authority and direction by a properly designated commander over assigned and attached forces in the accomplishment of the mission. Command and control functions are performed through an arrangement of personnel, equipment, communications, facilities, and procedures employed by a commander in planning, directing, coordinating, and controlling forces and operations in the accomplishment of the mission.

Command—The authority that a commander in the Armed Forces lawfully exercises over subordinates by virtue of rank or assignment. Command includes the authority and responsibility for effectively using available resources and for planning the employment of, organizing, directing, coordinating, and controlling military forces for the accomplishment of assigned missions.

Computing and communications—Two pervasive enabling technologies that support C2 and intelligence, surveillance, and reconnaissance. Computers and communications process and transport information.

Control—Authority which may be less than full command exercised by a commander over part of the activities of subordinate or other organizations. Physical or psychological pressures exerted with the intent to assure that an agent or group will respond as directed.

Intelligence (I)—The product resulting from the collection, processing, integration, analysis, evaluation, and interpretation of available information concerning foreign countries or areas. Information and knowledge about an adversary obtained through observation, investigation, analysis, or understanding.

Sometimes the term "C4ISR" is employed. The additional elements included in C4ISR are the following:

Surveillance—The systematic observation of aerospace, surface or subsurface areas, places, persons, or things, by visual, aural, electronic, photographic, or other means.

Reconnaissance—A mission undertaken to obtain, by visual observation or other detection methods, information about the activities and resources of an enemy or potential enemy, or to secure data concerning the meteorological, hydrographic, or geographic characteristics of a particular area.

Two additional terms are commonly used in describing C4I capabilities:

Situational awareness—The knowledge of where you are, where other friendly elements are located, and the status, state, and location of the enemy.

continues

> **Information superiority**—The relative advantage of one opponent over another in commanding and controlling his force. Information superiority or dominance is achieved both through the training of leaders to make rapid and appropriate decisions using superior technical information means provided to them, and through efforts to degrade and deny these same capabilities to an opponent while protecting one's own capability.
>
> SOURCE: Joint Chiefs of Staff, *Department of Defense Dictionary of Military and Associated Terms*, as amended through December 7, 1998 (Joint Publication 1-02).

judgment—including difficult-to-quantify aspects of human behavior (such as fatigue, experience level, and stress), the uncertainty of data, and the plausible future states resulting from actions by both their own force and the enemy—to make decisions about future actions and how to convey those decisions in ways to facilitate their proper execution. In doing so, commanders are supported by tools to enable and accelerate the planning and decision-making process, to achieve the decision-making superiority envisioned by DOD. And, of course, to be effective, command decisions must be implemented, a process to which C4I technologies are also relevant (e.g., in speeding up the link through which targeting information is passed to weapons, the so-called sensor-to-shooter link). The development and use of the right tools allow the commander to focus better on those issues associated with the essence of command—the art versus the science. As more and better-automated tools are developed and people are trained to use them, it will become even more important to recognize the art of command as distinguished from the mechanics of the tools used to provide information.

1.2 THE IMPACT OF C4I ON MILITARY EFFECTIVENESS

1.2.1 Evidence from Recent Experience

Although the Gulf War was plagued by innumerable problems with C4I capability, timeliness, and interoperability among both U.S. and allied forces, the real-world impact of C4I technology in enhancing the effectiveness and security of the coalition forces was amply demonstrated.

The C4I capabilities on which allied forces depended were highly tenuous and relied on inadequate methods for construction and distribution of operational plans and execution orders (e.g., the air tasking order had to be delivered manually to ships at sea), collection and assessment of battle damage information, and coordination of operations on a global

> **BOX 1.2 Examples of C4I Systems**
>
> The following examples of some current C4I systems are intended as illustrative only; many other C4I systems would serve equally well to provide context and orient the reader to the myriad of C4I systems used currently by DOD.
>
> - *Global Command and Control System.* The Global Command and Control System is designed to provide an integrated picture of the battlefield as well as core planning and assessment tools required by combatant commanders and joint task force commanders. The system includes a growing set of applications including (1) the Joint Operational Planning and Execution System, which is used to plan and execute joint military operations, and (2) the Requirements Development and Analysis application, which generates the time-phased force and deployment database for an operation (including time-phased force, non-unit-related cargo, and personnel data; data on movement for the operation plan; units to be deployed to support the operation plan; routing of forces to be deployed; data on movement associated with deploying forces; and transportation requirements).
> - *Contingency Theater Air Planning System.* The Contingency Theater Air Planning System assists theater-level air battle staffs with the development and execution of air tasking orders, which lay out the strike plan for air assets and control the operation of all other airborne assets.
> - *Joint Maritime Command Information System.* The Joint Maritime Command Information System is the Navy's designated command and control system for the future Global Command and Control System. It supports command and control and tactical intelligence warfighting requirements for afloat, ashore, and tactical/mobile units. The Joint Maritime Command Information System provides timely, accurate, and complete all-source C4ISR information management and develops a common operational picture for warfare mission assessment, planning, and execution. It incorporates the Marine air-ground task force C4I software.
> - *Maneuver Control System.* The Maneuver Control System provides units with the multidimensional (air, land, sea, and space) order of battle and rules of engagement. For example, it provides Army tactical commanders and their staffs (corps through battalion) automated, online, near-real-time systems for planning, coordinating, and controlling tactical operations.
> - *Advanced Field Artillery Tactical Data System.* This system provides automated fire support command and control functions, including tactical fire direction, fire planning, fire mission execution, and fire asset control.
> - *Joint Tactical Information Distribution System.* The Joint Tactical Information Distribution System provides secure, anti-jam-protected digital data and voice communications for theater, air, ground, and naval forces. The system is designed to enhance combat capability in fighter aircraft, command and control platforms, and surface air defense units, and it provides a data transfer link between weapon platforms and C4I systems for real-time situation awareness, targeting, and mutual support.

> **BOX 1.3 Major Elements of the
> Defense Information Infrastructure**
>
> *Defense Information Systems Network*
>
> The Defense Information Systems Network is the global, end-to-end, information transfer infrastructure of the DOD. The Defense Information Systems Network provides the communications infrastructure and services needed to satisfy national defense command, control, communications, and intelligence requirements and meet worldwide U.S. defense requirements. The purpose of the Defense Information Systems Network is to enable rapid, reliable and secure information access to conduct effective military operations, and, in particular, to allow any warrior to perform any mission, any time, any place in the world, based on information needs. The network's architecture prescribes a global network integrating DOD-wide communications systems assets, military satellite communications, commercial satellite communications initiatives, leased telecommunications services, dedicated DOD service and defense agency networks, and mobile/deployable networks, i.e., the consolidated worldwide enterprise-level telecommunications infrastructure that provides the end-to-end information transfer component of the Defense Information Infrastructure (DII).
>
> The Defense Information Systems Network infrastructure consists of the sustaining base (i.e., base/post/camp/station) C4I infrastructure (including legacy systems) that interfaces with the long-haul network in order to support the deployed warfighter with reach-back services, the long-haul telecommunications infrastructure (including today's defense communications systems and the communication systems and services between the fixed environment and the deployed (joint task force/combined task force) warfighter), and the deployed warfighter and associated telecommunications infrastructures that support the joint task force/combined task force.
>
> *The Defense Information Infrastructure Common Operating Environment*
>
> The Defense Information Infrastructure Common Operating Environment (DII-COE) is a software infrastructure for supporting DOD's C3I and combat support applications. It consists of a collection of reusable software components (commercial off the shelf (COTS) and government off the shelf) along with a set of guidelines, applications program interfaces, and built-in conformance with standards specified in the Joint Technical Architecture.
>
> The key goals of the DII-COE are interoperability among joint service applications and data, software reuse, and rapid information retrieval. The payoff of a common software infrastructure lies in the reduction of costs related to acquisition, operations, and support. Acquisition costs can be reduced by taking advantage of commercial trends and COTS software products. Reductions in operations and support costs will be attained with government off-the-shelf software reuse, easier system upgrades to new software
>
> *continues*

versions or platforms, and a common environment for operations and training.

The software structure of the DII-COE is composed of three layers: the kernel, infrastructure services, and common support applications. The kernel consists of the computer's operating system (e.g., Solaris, HP/UX, Windows NT, etc.) and fundamental services for desktop functions (e.g., display presentation, file management, printing, and network and system administration). The infrastructure services layer contains utilities, tools, and software for network and database management (e.g., relational database server/tools), and communications and presentation services (e.g., TCP/IP, World Wide Web browser, etc.). The common support applications layer contains software for message processing (e.g., Automated Message Handling system, map display development via the Joint Mapping Tool Kit, track correlation, alerts, help, and office automation).

The Defense Message System

The Defense Message System is a joint DOD program created to improve the department's electronic messaging capabilities while reducing the cost associated with the current messaging systems. The Defense Message System is undergoing an evolutionary transition from the baseline Automatic Digital Network and electronic messaging services to an integrated system using the common user communications transport provided by the Defense Information Systems Network. During the transition, the Defense Message System requires the ability to maintain interoperability between the baseline systems, the allied messaging systems, other governmental agencies, and commercial messaging users. The target Defense Message System is based on international standards for messaging, directory, and service management. It will employ security services as approved by the National Security Agency to provide protection appropriate to the required level of trust.

C3 and Combat Support Applications

Global Command and Control System. The Global Command and Control System is intended to provide combatant commanders one integrated resource for generating, receiving, sharing, and using information securely. It provides for surveillance and reconnaissance information and access to global intelligence sources as well as data on the precise location of friendly forces. The Global Command and Control System provides support for crisis planning, intelligence analysis, tactical planning and tactical execution, and collaborative planning. It establishes the top-level technical infrastructure for automated support to command and control (C2) operations. The Global Command and Control System supports the National Command Authorities and subordinate elements in the generation and application of national power. It is intended to provide for maintenance of a common perception of the crisis or battlespace, access to planning support information, collaborative access to a common operational plan, visibility of plan execution status,

continues

and adaptive control of communication and information centers for surge needs and users with degraded communications.

Global Combat Support System. The Global Combat Support System provides information access and fusion across the entire spectrum of combat support. The Global Combat Support System provides each combat support functional area—supply, transportation, finance, medical, personnel, acquisition—with access to authoritative data and integrating existing combat support information to gain efficiency and interoperability in support of the warfighter. It is designed to overcome existing shortfalls in the limited breadth of isolated and stovepiped systems by combining and/or fusing data provided from multiple authoritative sources into relevant, coherent, integrated information. It applies current information technology to provide that full spectrum of information system capabilities to the warfighter and to the sustaining bases. The Global Combat Support System will enable accurate and real-time combat support information to be available to the National Command Authorities, services, CINCs, the joint task force commanders, and service components. The Global Combat Support System is a demand-driven, joint warfighter-focused capability to accelerate delivery of improved combat support effectiveness.

Ultimately, both the Global Combat Support System and the Global Command and Control System applications will be available on the same workstation to provide a truly integrated view of the battlespace.

Theater Deployable Systems

The Standardized Tactical Entry Points program provides global access to standardized Defense Information Systems Network services that support deployed joint task forces. Standardized Tactical Entry Points constitute a global network that provides interoperable communications between the strategic and tactical forces and provides essential circuits and worldwide information transfer capability by using the Defense Information Systems Network. The Standardized Tactical Entry Points network provides standard/prepositioned C4I communications for the warfighter and improves tactical access to strategic voice and data services, tactical/strategic communications interoperability, deployed tactical commanders' access to headquarters, CINCs, and the Pentagon, and interoperability and reach-back for the tri-service tactical ground mobile forces and the Navy shipboard tactical users.

Network and System Management

Management of the DII as a whole is performed by a combination of the Defense Information Systems Agency and the CINCs, services, and agencies that work in collaboration to provide an end-to-end enterprise view of the DII. This collaboration of systems, roles, and responsibilities is termed the

continues

> Joint DII Control System. From a network and systems management perspective, the DII is composed of three "blocks" or domains: the sustaining base block (managed locally by CINC/service/agency control), the long-haul block (managed by the Defense Information Systems Agency), and the deployed block (managed by the joint task force commander). The Joint DII Control System establishes the operational integration of the systems and network management roles, responsibilities, and relationships across all three "blocks" or domains of the DII. It will also result in establishing the common operating picture that will be shared by the CINC/service/agency managers.
>
> The Joint DII Control System is based on a jointly defined technical architecture, interface standards, and performance standards derived from the Joint Technical Architecture and the DII-COE. The ultimate goal of the Joint DII Control System is to field a capability whereby all DII users and providers will be able to share a common picture of their DII assets and supporting infrastructure. The Joint DII Control System will also provide a converged capability with information assurance and defensive information operations to ensure that a fully articulated picture is available for global situational awareness.

scale among systems ranging from highly sophisticated to significantly outdated.

Nonetheless, given sufficient time (in the case of the Gulf War, nearly 6 months) to prepare, a formidable capability was established for command and control of a multinational force in a region of the world where virtually no infrastructure previously existed to accommodate such complex operations. C4I has been reported in numerous after-action media as a major force multiplier in the conflict. For example:

- C4I systems supported—through simultaneous suppression of enemy air defenses—highly effective, precise, orchestrated strikes on a variety of targets in Baghdad on the initial night of war, with extremely low casualties.
- The Global Positioning System allowed orchestrated movements of coalition armored forces to outflank Iraqi forces and engage them at the maximum effective range of coalition weapons.

More recently in Bosnia, advanced C4I technology has provided forces with enhanced capabilities to detect, process, decide, and communicate. For example:

- The Predator Unmanned Aerial Vehicle has improved monitoring of compliance with the Dayton Peace Accord.

- Linked Operations-Intelligence Centers Europe[4] systems have facilitated the sharing of intelligence among selected coalition partners.
- The Joint Surveillance Target Airborne Radar System supported the insertion of ground forces into Bosnia.

Many warfighters involved came away from the Gulf War with the view that improving C4I capability and interoperability would add more to military operations than additional improvements in weapons. Continued improvement in the precision and/or lethality of weapons remains a priority; in fact, such enhancements in capabilities may well result more from application of C4I improvements than from near-term advancements in weapons technology. In addition, the challenges of operating in urban environments and in rough, wooded areas must be addressed rather than simply extrapolating the successes achieved in a desert environment.

1.2.2 Potential Impact of C4I on Military Operations

The examples below are illustrative of how many military thinkers conceptualize the potential impact of C4I on military operations. Some evidence to support these concepts is available from studies and exercises and experiments,[5] but for the most part their full significance has not been demonstrated in real-life operational scenarios.

Information Superiority and Greater Situational Awareness

To exercise authority and direction effectively in combat and other military operations, commanders must have situational awareness. Use of information technology to make a commander's situational awareness better also creates the potential to improve the effectiveness with which the commander directs and controls his forces. To the extent that the promise of C4I technologies is realized, reduced force size might be compensated for by information superiority—the ability of a force to have, and protect, a comprehensive view of enemy and friendly forces as well as the combat environment, while denying the enemy a comparable capa-

---

[4]Linked Operations-Intelligence Centers Europe is the U.S. European Command's system that provides U.S. and NATO forces, and other allied forces, with near-real-time correlated situation and order-of-battle information. For more information see Joint Distributive Intelligence Support System Program Office, online at <http://www.jdisspmo.org/relpro/loce.htm>.

[5]See, for example, H.S. Marsh and P.J. Walsh, *Employment Strategies and CONOPS Enabled: A Compilation of Draft White Papers on Future Employment Strategies and CONOPS Enabled Prepared to Support the C4ISR Mission Assessment*, November 22, 1996, Draft, SRI International.

bility—where it can be shown that information superiority is a force multiplier.

The growing list of land-, air-, sea-, and space-based sensors combined with other sources makes the fusion of information an essential dimension of situational awareness. Fusion of data from this multitude of sources is indispensable to achieving information superiority in the regional environment. The challenge in doing so goes beyond the receipt and display of sensor data to include reconciling those data (eliminating redundancy and outdated information) and extends to the fusion of multiple sources of information into timely and meaningful intelligence. Through this process, true information dominance can be achieved. In that regard, information dominance must also include situational awareness with regard to space-based systems. Knowing friendly, enemy, and neutral satellite coverage and capability will be of vital concern to the joint commander and his component commands.

The cornerstone of information superiority is advanced C4I technology and systems, which can provide to all tactical levels of command a robust, continuous, common operating picture of the battlespace.[6] The resulting heightened situational awareness should vastly improve the effectiveness with which commanders at all levels can pursue a mission. The common operating picture can allow tactical decision making at the lowest levels of command consistent with the higher-level commander's operational objectives, and the decentralized tactical execution can enhance the ability of lower-level tactical units to react quickly to changing circumstances. A common operating picture is a central element in a number of initiatives, including the following four:[7]

- *The Army Digitization Master Plan (Force XXI).* The Army Digitization Master Plan is intended to "create a simultaneous, common picture of the battlefield from soldier to commander at each echelon

---

[6]In some usages, the term "common operating picture" refers to a view of the battlespace that is near-real-time; in other usages, it refers to a view that lags by as much as an hour. This report adopts neither usage, preferring instead to make the time dimension explicit when it is relevant to the discussion.

[7]This is not to say that the notion of a common operating picture is new. For example, the foundation of the Navy's Joint Maritime Command Information System is a common operating picture of a battlespace that is relevant to Navy operations, and JMCIS has been in existence since around 1993. The JMCIS common operating picture integrates reports from a variety of sensors, including some on the ship where the common operating picture is displayed and other off-board sensors on accompanying platforms dispersed in the battlespace. However, because JMCIS is oriented toward Navy operations, the JMCIS common operating picture is available primarily to surface and subsurface platforms. The intent of the programs described is to pass a common operating picture to tactical echelons that are much lower in the command hierarchy.

through the networking of sensors, command posts, processors, and weapons platforms. The program provides the communication and displays which allow participants to aggregate information and maintain an awareness of what is happening around them, both friendly and enemy. Digitization allows the employment of forces in a highly mobile, synergistic, and overwhelming manner."[8] Warfighting experiments, designed to test the concept of digital command and control, suggest that the Army may be able to significantly reduce the size of its mechanized division while increasing the physical space for which it is responsible in a traditional conflict.

- *The Theater Air and Missile Defense program.* Conducted by the Ballistic Missile Defense Organization, the Theater Air and Missile Defense program seeks to develop capabilities to display a single integrated air picture, available to all relevant units in the theater, that is accurate, resolved, consistent, and timely (essentially real-time). The single integrated air picture, which would integrate data on air (and cruise-missile) threats provided by multiple sensors (possibly of different types) and sources located on different platforms, is intended for use by commanders at all levels to identify, prioritize, and execute air defense engagements.

- *The Battlefield Awareness and Data Dissemination (BADD) advanced concept technology demonstration (ACTD).* The purpose of the Battlefield Awareness and Data Dissemination ACTD, which is supported by the Defense Advanced Research Projects Agency, is to develop, install, and evaluate an operational system that would allow commanders to design their own information system; deliver to warfighters an accurate, timely, and consistent picture of the joint/coalition battlefield; and provide access to worldwide data repositories. It integrates a wideband, low-cost broadcast mechanism, information management services providing user access to a wide variety of information sources (including unmanned aerial vehicle and national imagery; Global Command and Control System operational data; and combat information systems such as the U.S. Army's All Source Analysis System, the Joint Maritime Combat Information System, the U.S. Air Force Combat Intelligence System, and the Common Ground Station), and battlefield awareness services that present to the user a coherent picture of enemy and friendly forces integrated with terrain, image, and video data.[9]

---

[8]Army Digitization Office. 1996. *Army Digitization Master Plan, 1996*, Army Digitization Office, Washington, D.C., March.

[9]Adapted from the BADD program overview: Office of the Under Secretary for Acquisition and Technology, 1999, *Battlefield Awareness and Data Dissemination*, Office of the Under Secretary for Acquisition and Technology, Department of Defense, Washington, D.C.; available online from <http://www.acq.osd.mil/at/badd.htm>.

- *The "Extending the Littoral Battlespace" (ELB) ACTD.* The ELB ACTD is intended to enhance the advanced warfighting concepts of the Navy and Marine Corps by providing or enabling theater-wide situation awareness, integrated sensors, responsive remote fires and targeting, and over-the-horizon connectivity. Further, it proposes a range of operational and tactical concepts that leverage command, control, communications, computational, and other technologies to exploit information and improve precision firing and targeting in future operations. For example, it would enable the effective employment of dispersed and disaggregated units as well as increasing the capability for rapid operations by conventionally configured forces. Disaggregated units could operate in an enlarged battlespace, presenting few concentrated targets to the enemy while employing massed remote firepower to harass, damage, and destroy. Central to ELB is a beyond-line-of-sight tactical information infrastructure with wideband communications networks and enhanced situation understanding that would provide common situational awareness at all levels of command.[10]

Decentralized Freedom of Action

The transmission of a common operating picture to each unit, in real time and in parallel, would enable commanders at the tactical level to quickly grasp the larger battle picture and thus to determine local unit objectives with much greater latitude and assurance.[11] Also, lower echelons of command could quickly orient their units to new orders and specific objectives and pursue those objectives with greater freedom of action, all within the framework of the overall objectives of the joint or combined force commander. Peer unit collaboration in achieving local objectives and increased autonomy would thus become more feasible and could lead to higher operational tempos.

Using C4I to Conduct Precision Strikes

In the traditional context of ground warfare, overwhelming force was applied by massing forces at points of contact with the enemy.

---

[10]Adapted from the ELB program overview: Office of the Under Secretary for Acquisition and Technology, 1999, *Extending the Littoral Battlespace*, Office of the Under Secretary for Acquisition and Technology, Department of Defense, Washington, D.C.; available online from <http://www.acq.osd.mil/at/eld.htm>.

[11]A potential downside to a common operating picture is that detailed awareness at *all* levels of command above those that are the "trigger pullers" creates the potential for second-guessing, with a negative impact on the initiative of those who are engaged in combat. Whether this and other potential problems in fact turn out to be real problems, and if they are, how they can be managed, are research areas that need to be explored.

Through the use of modern C4I technology and systems, together with "smart" weapons that are guided directly to their targets, massed effects without massed forces can indeed be possible. Indirect fire launched from assets that are widely dispersed and not in direct contact with enemy forces could produce effects comparable to those possible with forces massed at points of contact. Sensors would be deployed close to or within enemy operating areas and would be linked directly to the forces that are engaging those enemy forces. These same sensors would be used to feed the C4I infrastructure that provides the real-time common operating picture of the battlespace. Such an approach would significantly increase the effectiveness of remotely delivered firepower, reduce friendly losses, and provide significant increases in the effectiveness of the maneuver forces, thus constituting a major shift from the traditional notion of attrition-based warfare. Box 1.4 describes how a current ACTD may be able to reduce the cycle time for striking time-critical targets.

Using C4I to Enhance the Effectiveness of Air Operations

Advanced C4I offers the means to achieve greatly improved effectiveness in carrying out most of the challenging tasks in air operations. The single integrated air picture is critical to improving the effectiveness of the air and missile defense missions. Creating a single integrated air picture (SIAP) is a significant technical challenge, given the extremely short time lines against which the air assets must operate.

The United States has been striving for a single integrated air picture since 1969 when the Tactical Air Control System/Tactical Air Defense System program was launched. Subsequent developments have yielded capabilities that allow the creation and maintenance of a single integrated air picture, but these systems still have clear deficiencies in such areas as integration and the ability to share information with potential coalition partners.

Today the problem has become even more challenging with increased concerns about ballistic missiles and stealthy cruise missiles. Engaging targets that are mobile or relocatable or that have short dwell times is another challenging air task that could be improved through rapid assessment of target changes and feedback to the attacking units. The use of advanced sensor technology and the fusion of data from applicable sensors of each of the services could help the further development of a single integrated air picture. This advanced C4I capability would also enhance the effectiveness of precision guided weapons against fixed targets by providing timely and precise target location information. Box 1.5 describes how networked sensors can improve air defense.

BOX 1.4 Reducing the Cycle Time for Striking Time-Critical Targets

Emergent targets are high-payoff land and maritime platforms, force groupings, and geographic complexes that must be attacked inside cycle times that are not consistently achievable by the current Joint Targeting Process. The effective attack of such targets demands a seamless flow of information across service, organization, and system boundaries if they are to be consistently attacked within their short windows of vulnerability (1 to 2 hours or less). Emergent targets operate inside the Joint Targeting Cycle because information is decelerated as it crosses organizational and system boundaries. Numerous studies have documented the latency introduced into the targeting process when data has to be rekeyed, air gapped, disseminated in hard copy, or otherwise transferred manually between systems; and even when such impediments do not exist one component of the force rarely has visibility into what strike assets the other has available against what portions of the battlespace on a time-sensitive basis—so that requests for other services' weapons are made inefficiently or not at all. The warfighter is not receiving the full benefit of our massive investment in information and weapons technology.

The Joint Continuous Strike Environment advanced concept technology demonstration seeks to improve the responsiveness of U.S. strike cycles against emergent, time-critical surface targets. The Joint Continuous Strike Environment functionality will encompass deep-strike assets from all services and selected allied assets. It will take advantage of existing but untapped potential for servicing emergent targets to shunt and accelerate information along the sensor-to-shooter pathways, thus enabling a joint force commander to hold emergent targets at risk without disrupting other aspects of his campaign plan. Whether a target pops up due to enemy action, or emerges because it is critical to accomplishing a joint force commander's plans in a temporally dominated battlespace, the Joint Continuous Strike Environment will provide the tools to put the right weapon on the right target at the right time.

Its goal is to reduce by at least one order of magnitude the latency associated with correlating command guidance, weapons, targets, and airspace deconfliction and launching attacks against emergent targets. The Joint Continuous Strike Environment provides to warfighters automated target prioritization based on a commander's guidance and objectives, continuous monitoring of weapon availability (resulting in visibility into all service weapons rather than today's service-centric view at execution nodes), optimized weapon target pairing in which actionable intelligence is matched to available strike assets, and near-real-time airspace deconfliction that avoids the need to constrain operations throughout the theater by altitude and volume.

SOURCE: Adapted from the description of the Joint Communications Strike Environment program available online at <http://www.cisa.osd.mil/hostedsites/jcse/overview.htm>.

BOX 1.5 Network of Sensors Approach to Theater Air Defense Against Cruise Missiles

Fast, low-flying cruise missiles attacking targets on land or at sea are a very difficult threat against which to defend. In principle, engagements of an incoming cruise missile far away from the threatened target are highly desirable, because such engagements allow multiple attempts to destroy the cruise missile. (An important collateral benefit is that the long-range destruction of a cruise missile carrying chemical or biological weapons reduces the likelihood that the chemical or biological weapons agent will affect the target.)

Cruise missiles can be engaged with surface-to-air missiles or fighters. In the case of a surface-to-air missile engagement, the range at which it occurs is limited by one of two factors—the fly-out range of the missile itself and the range of the sensors (usually radar) that guide it to the target. However, the range of a ground-based radar is limited by the line of sight to the horizon, which is typically much smaller than the missile's fly-out range.

The horizon line-of-sight limitation can be overcome by increasing the altitude of the radar (e.g., placing it on an airborne platform) and thus increasing the radar line-of-sight range to the horizon, or by using over-the-horizon sensors to guide the missile. It is often the case that over-the-horizon sensors are present, but in general these sensors will be associated with platforms other than the one that can fire the surface-to-air missile.

In any event, a network of sensors providing the right kinds of data can in principle support surface-to-air missile engagements for any surface-to-air missile within fly-out range of its target. Since fly-out ranges are often four to five times the distance to the radar horizon, the improvement in air defense coverage is significant.

Today, the ability to employ networks of heterogeneous sensors is limited by the fact that fire-control-quality data cannot in general be shared among all the shooters that might come into play in an engagement. Moreover, the "stovepipe" architecture in place can prevent even the surveillance data generated by some sensors from being available to certain shooters. The Navy's program to develop the Cooperative Engagement Capability system is intended to provide such functionality for air defense over water; similar developments are under way to provide comparable capabilities over land.

SOURCE: Adapted from Joint C4ISR Decision Support Center. 1997. *Precision Engagement C4I Operational Architecture Study (Sensor-to-Shooter III)*, Joint C4ISR Decision Support Center, Office of the Assistant Secretary of Defense for Command, Control, Communications, and Intelligence, Department of Defense, Washington, D.C.

1.3 THE U.S. MILITARY'S WORK IN EXPLOITING INFORMATION TECHNOLOGY

The Joint Chiefs of Staff and the military services have taken note, in a number of studies, of the role of information technology in future military operations. In particular, DOD has identified a technology-enabled "revolution in military affairs" as one that involves "harnessing new technologies to give U.S. forces greater military capabilities through advanced concepts, doctrine, and organizations so that they can dominate any future battlefield."[12] Joint Vision 2010 is based on four broad operational concepts: dominant maneuver, precision engagement, full-dimension protection, and focused logistics (Box 1.6). For each of these concepts, information superiority is a critical enabler.

Each of the services is exploring the implications of Joint Vision 2010 for itself, taking steps with experimental studies, wargames, research and development investments, advanced concept technology demonstrations, and simulation gaming to develop and test concepts and capabilities that will ensure military preparedness for the 21st century. The goal is to understand how to more effectively organize, equip, and train military forces. The effort goes far beyond learning how to modernize current weapons systems, and includes how to deploy and employ new systems, and how to support these systems efficiently and effectively at a lower cost and within a drastically reduced cycle time. (Box 1.7 describes service initiatives in more detail.) Additionally, as an extension of individual service experimentation, and in response to congressional pressures, a joint experimentation activity is being established at the U.S. Atlantic Command to address the co-evolution of doctrine, tactics, and new technological capabilities.

1.4 THE ROLE OF C4I SYSTEMS IN FUTURE MILITARY ENVIRONMENTS

1.4.1 Likely Environments of Future Military Operations

The 21st century will see the U.S. military continuing to be fully committed to responding to a full spectrum of missions, from peacekeeping and other military operations other than war to major theater war. These operations will be conducted in a world where sophisticated military equipment can be purchased by anyone with adequate funds, and some military capabilities can be purchased through commercial markets. Com-

---

[12]William S. Cohen. 1998. *Annual Report to the President and to Congress*, Department of Defense, Washington, D.C., Chapter 1.

> **BOX 1.6 Joint Vision 2010**
>
> Joint Vision 2010 provides a "conceptual template" for the improved conduct of joint warfighting operations by leveraging technological advances. Joint Vision 2010 stresses the importance of information superiority—defined as "the capability to collect, process, and disseminate an uninterrupted flow of information while exploiting or denying an adversary's ability to do the same"—as the basis for improved command, control, and intelligence functions.
>
> It is based on four emerging operational concepts that taken together will allow the U.S. armed forces to "dominate the full range of operations from humanitarian assistance, through peace operations, up and into the highest intensity conflict (i.e., full spectrum dominance)":
>
> - *Dominant maneuver* is the "multidimensional application of information, engagement, and mobility capabilities to position and employ widely dispersed air, land, sea, and space forces to accomplish assigned operational tasks."
> - *Precision engagement* is a system of systems that allows U.S. armed forces "to locate the objective or target, provide responsive command and control, generate the desired effect, assess [their] level of success and retain the flexibility to reengage with precision when required."
> - *Full-dimension protection* provides multilayered protection for forces and facilities ranging from theater operations to the individual soldier.
> - *Focused logistics* is the "fusion of information, logistics, and transportation technologies to provide rapid crisis response, to track and shift assets even while en route, and to deliver tailored logistics packages and sustainment directly at the strategic, operational, and tactical level of operations."
>
> Joint Vision 2010 identifies several critical considerations necessary to implement these new operational concepts: high-quality personnel, innovative leadership, joint doctrine, joint education and training, agile organizations, and technology enhancements.
>
> SOURCE: Chairman of the Joint Chiefs. 1996. *Joint Vision 2010*, Joint Chiefs of Staff, Washington, D.C.

mercialization in such areas as information technology, space operations, imaging, and global positioning, and the increased need and desire of the United States to use commercial technology for military use, reduce the ability of the United States to protect these technologies. Also, when an adversary is able to make use of commercial space and information tech-

BOX 1.7 Service Initiatives to Leverage Information Technology

Army

The Army's contribution to joint operations is "the ability to conduct prompt and sustained operations on land throughout the entire spectrum of crisis." Army Vision 2010 also lays out a vision between the Army's ongoing and relatively near term (FY 04 and sooner) Force XXI implementation process, and the longer-term vision of the Army After Next, which looks at the future geostrategic environment (i.e., 30 years out). In the future, the Army plans to focus the execution of its responsibilities through "a deliberate set of patterns of operations": project the force, protect the force, shape the battlespace, conduct decisive operations, sustain the force, and gain information dominance. (The latter is fundamental to the five other patterns, as well as to the operational concepts of Joint Vision 2010.)

Both the Force XXI and the Army After Next processes are identifying new concepts of land warfare that have implications for the Army's organization, structure, operations, support, and materiel. Force XXI's premise is that greater situational awareness, obtained by leveraging information technology, particularly from the commercial sector, on current platforms (Abrams tanks, Bradley infantry fighting vehicles, and Apache helicopters) will provide friendly commanders with greater "mental agility" and thus increase the lethality, survivability, and operations tempo of their forces.

The Army's Experimental Force, the 4th Infantry Division at Ft. Hood, Texas, is the vehicle for testing these innovations. The Experimental Force is a heavy force used to identify and evaluate, through a series of Army advanced warfighting experiments, new operational concepts, organizational designs, doctrine, and tactics that take advantage of "digitization" technologies and the capabilities they offer. The Experimental Force also will examine flexible, highly tailorable organizations—from individuals to small units to echelons above corps—to meet the diverse needs of future operations.

At the same time, the premise of the more futuristic Army After Next is that greatly increased strategic and tactical mobility—i.e., physical agility—and all-encompassing "knowledge" of the battlespace—i.e., mental agility—will be the dominant factors in wars of the first quarter of the next century. As a result the Army is examining "leap ahead" technologies that will result in much lighter, smaller, more durable equipment that will enhance deployability and reduce the sustainability burden, while generating the lethality necessary for decisive operations.

Through an annual cycle of wargames, workshops, and conferences, the Army After Next strives to lay the research foundation necessary for assessing the effects of increased mobility, lethality, and maneuverability,

continues

and to ensure that land power remains a strategically decisive element of warfighting into the 21st century. The largest part of the effort is focused on examining the impact of technologies and system concepts for both air and land vehicles to provide significantly increased strategic and tactical mobility. From a command and control perspective, the goal will be to greatly facilitate the decision-making process for protecting, projecting, and employing the force. Use will be made of advanced, highly mobile, and easy-to-use sensors; communications; and processors that collect and distribute data throughout the battlespace, develop information, and create the knowledge to enable and ensure effective freedom of maneuver and dominant lethality. The innovations selected during this process will be tested by the 2nd Armored Cavalry Regiment at Ft. Polk, Louisiana.

(SOURCE: Department of the Army. 1996. *Army Vision 2010*, Department of the Army, Washington, D.C.)

Air Force

The Air Force's future vision is given in *Global Engagement: A Vision for the 21st Century Air Force*. *Global Engagement* is a strategic plan for meeting the Air Force's challenge of dominating air and space as a unique dimension of military power in the 21st century. The Air Force identifies six core competencies—air and space superiority, global attack, rapid global mobility, precision engagement, information superiority, and agile combat support—and is committed to ensuring these components through innovation. Air and space superiority will allow all U.S. forces freedom from attack and freedom to attack, while the Air Force's ability to attack rapidly anywhere on the globe will continue to be critical. Rapid global mobility will help ensure that the United States can respond quickly and decisively to unexpected challenges to its interests. The Air Force's precision engagement core competency will enable it to reliably apply selective force against specific targets simultaneously to achieve desired effects with minimal risk and collateral damage. Air- and space-based assets will contribute to U.S. forces' information superiority, and agile combat support will allow combat commanders to improve the responsiveness, deployability, and sustainability of their forces.

To better understand the potential offered by advanced technologies, the Air Force conducted its Expeditionary Force Experiment in September 1998. In that scenario, a rogue nation attacked a U.S. ally that requested U.S. assistance in halting the invasion. An air expeditionary force was deployed in response, and the experiment tested the ability to exercise coherent command and control through the use of forward and rear (continental U.S.-based) joint air operations centers and to plan and execute combat missions en route to the area of hostility. Under the experiment scenario, a much smaller number of command and control military per-

continues

sonnel and much less equipment were deployed to the combat area, with the rear joint air operations center housing the bulk of these personnel and equipment (as well as the joint force air component commander).

In the area of information superiority, the Air Force will focus on future global battle management/command and control systems to allow for real-time control and execution of all air and space missions, exploit unmanned aerial vehicle technology (especially in intelligence, surveillance, and reconnaissance and communications applications), and expand its defensive information warfare efforts.

The Air Force has established six new battle laboratories to implement its vision. The mission of these battle labs is to identify and validate innovative ideas that improve the ability of the Air Force to execute both its core competencies and joint warfighting. The concepts validated in the labs will be assimilated into Air Force organizational, doctrinal, training, and acquisition efforts. The six labs are concentrating on the following areas: unmanned aerial vehicles, information warfare, air expeditionary forces, space capabilities, battle management command and control, and force protection.

(SOURCE: Department of the Air Force. 1996. *Global Engagement: A Vision for the 21st Century,* Department of the Air Force, Washington, D.C.; Air Force Experimentation Office *EFX Public Web Site,* available online at <http://efx.acc.af.mil>.)

Navy

The building blocks of forward-deployed Navy and Marine Corps forces that contribute to peacetime presence, crisis response, and regional conflicts are the air carrier battle groups and the amphibious ready groups, which are highly flexible formations. The naval services will focus on a new direction to "project power from the sea in the critical littoral regions of the world," and have committed to structuring their expeditionary forces so that they are inherently prepared for joint operations.

The Navy's future vision of warfare, delineated in *From the Sea* and *Forward...From the Sea,* and further developed in the *Navy Operational Concept,* identifies five fundamental roles for the Navy: sea control and maritime supremacy, power projection from sea to land, strategic deterrence, strategic sealift, and forward naval presence. However, in the future the Navy will fulfill these roles with enhanced capabilities. The Navy has embraced a concept called network-centric warfare: the ability of widely dispersed but robustly networked sensors, command centers, and forces to have significantly enhanced massed effects. Combining forward presence with network-centric combat power, the Navy intends to close time lines, decisively alter initial conditions, and seek to head off undesired events before they start. The naval contribution to dominant maneuver will use the sea to gain advantage over the enemy, while naval precision engage-

continues

ments will use sensors, information systems, precisely targeted weapons, and agile, lethal forces to attack key targets. Naval full-dimensional protection will address the full spectrum of threats, providing information superiority, air and maritime superiority, theater air and missile defense, and delivery of naval firepower. Finally, naval forces will be increasingly called upon to provide sea-based focused logistics for joint operations in the littorals.

The At-Sea Fleet Battle Experiments, to be overseen by the Maritime Battle Center, are designed to explore new concepts and emerging systems like the Maritime Fire Support Demonstrator, Cooperative Engagement Capability, and theater ballistic missile defense to evaluate their effects on fleet capabilities and determine future requirements. These experiments are limited in number to maintain their quality and are combined with other fleet exercises to maximize participation. The first of these experiments, Fleet Battle Experiment Alpha (conducted off southern California in March 1997), evaluated C4ISR capabilities, requirements for a sea-based combined joint task force, and other emerging concepts.

(SOURCE: Department of the Navy. 1996. *Forward...From the Sea*, Department of the Navy, Washington, D.C.)

Marine Corps

The Marine Corps strategy *Operational Maneuver from the Sea* foresees warfare that requires tactically adaptive, technologically agile, opportunistic, and exploitative forces. Individuals and forces must be able to rapidly reorganize and reorient across a broad range of new tasks and missions in fluid operational environments. The Marines will still need to project power ashore for a variety of potential tasks ranging from disaster relief to high-intensity combat.

This vision calls for the following actions: focus on an operational objective; use the sea as maneuver space; generate overwhelming tempo and momentum; pit strengths against weakness; emphasize intelligence, deceptions, and flexibility; and integrate all organic, joint, and combined assets. In order to implement these principles, new operational directions will be needed to enhance the integration of naval expeditionary forces, revolutionize forcible entry operations between the land and sea, and expand maritime maneuver across the spectrum of conflict. The vision also calls for modernizing capabilities in the following areas by capitalizing on new technology and approaches to doctrine, organization, tactics, and training: mobility, intelligence, command and control, fire support, aviation, mine countermeasures, and combat service support.

The focus of the Marine Corps Revolution in Military Affairs efforts is on the enhancement of the individual Marine and his or her ability to win in combat. The Marine Corps Combat Development System focuses on gen-

continues

erating the most effective combination of innovative operational concepts, new organizational structures, and emerging technologies. Through the 5-year "Sea Dragon" program, the Marines have developed an extensive experimentation plan divided into three phases, each culminating in an Advanced Warfighting Experiment:

- Hunter Warrior—designed to examine naval power projection in a dispersed, non-contiguous littoral battlespace, enhanced firepower and targeting, and C4I and the "single battle." The scenario established a situation in which naval forward-presence forces were tasked with conducting advance force operations in support of a friendly nation against invasion from a hostile neighbor, pending arrival of follow-on U.S. land-based forces. The experiment tested how doctrine, organization, training, equipment, and sustainment can be improved to produce the needed capabilities to implement new warfighting concepts.
- Urban Warrior—a 2-year effort, begun in 1997, to explore operations in urban, near-urban, and close terrain.
- Capable Warrior—combining virtual and live forces in operational-level deception and maneuver in response to crisis, with the objective of containing or preventing an incipient major theater war.

(SOURCE: U.S. Marine Corps. 1996. *Operational Maneuver from the Sea*, Headquarters, Marine Corps, Washington, D.C.)

---

nologies, it will be more difficult for the United States to preclude their use in time of conflict.

Given the U.S. military strengths and vulnerabilities and the difficulty, if not impossibility, of an adversary effectively matching the United States in organization, training, and military equipment, a potential adversary's strategy is likely to entail the development of asymmetric capabilities to effectively counter the United States. Asymmetric opportunities for a would-be adversary include finding low-cost means of precluding the U.S. ability to project its military power, particularly in landing forces in another country, by exploiting the aversion of the U.S. public for casualties; developing ways to counter the effectiveness of U.S. air power and precision munitions; and seeking ways to preclude or undermine U.S. information superiority.

Such trends portend a future in which low-cost ballistic and cruise missiles, weapons of mass destruction, and information attacks are a threat. Weapons of mass destruction, particularly chemical and biological weapons, will be available to the full range of threats, from rogue nations to transnational actors, international criminals, and terrorists. Attacks on targets within the continental United States may be launched to reduce the DOD's ability to command, control, deploy, and support its forces, or—if launched against non-military targets in the United States—to influence the American public. It will be harder to predetermine threats to U.S. interests, and attacks against the continental United States may well occur in the United States as terrorist attacks or as integral parts of an overall campaign against the United States.

The changed and changing world environment has a number of important military implications for the U.S. military. U.S. command and control must be global, capable of supporting a wide range of operations anywhere in the world, must operate in any terrain and on the move (by ship, plane, or land vehicle), and must be sustained from early warning and crisis management through post-conflict tasks. Also, given the U.S. public's aversion to U.S. military casualties, the U.S. military has placed an even greater emphasis on high-technology solutions, such as precision munitions and remote delivery.

Service component forces will operate jointly under a joint commander and, in many cases, will be combined with allied and coalition forces. To carry out command and control, the joint commander must receive information about the threat, operational environment, and status of his service component forces, and must be able to communicate with his component commanders about decisions related to the integrated allocation and employment of service assets.

As the United States responds to situations around the world, it will do so with other international powers, either regional allies or coalitions formed in response to the specific crisis, and operations will not only be joint, but combined. The type of missions and the international composition of the force will require coordination with multiple departments, agencies, and organizations (non-governmental as well as governmental), including those of coalition partners. The combined joint task force commander, when American, would have the same command and control requirements with his entire combined forces as he would have with his U.S. forces.

1.4.2 Rapid Planning to Support Rapid Response

Given the range of potential adversaries and the unpredictability of events that might challenge the interests of the United States, the need to consider the use of U.S. military forces could occur at any time. Despite the best available intelligence information, surprises will occur, and it is likely that there will be only a very short time period between indications of trouble and force employment, thus making rapid planning tools an essential C4I requirement at both joint command and service component command levels. Such planning tools can assist in determining force composition, force deployment, and probable battle outcomes. The planning process must be sufficiently flexible to accommodate situational changes as they unfold even as deployments are under way, as occurred during the deployment to Haiti. In that regard, rehearsal tools capable of receiving current intelligence, updating an original plan, and disseminating appropriate changes must be an integral part of the rapid planning process and must be made available to deploying units and their leaders. The tools that support planning and rehearsal must be able to run much faster than real time to explore the impact of alternative courses of action, and also to run at slower than real-time speeds to support rehearsal and learning.

The increase in operational tempo and the range of weapons employed demand that planning and execution be continuous, and not discrete, time-phased, sequential actions. As stated in *Joint Vision 2010*, "Real-time information will likely drive parallel, not sequential, planning and real-time, not prearranged, decision making."[13] Mobile communications and computers supporting command and control must be able to support operations en route on the land, at sea, and in the air. Command posts must be small, agile, and mobile to survive and remain relevant. How small can a command post be made, how can it be made redundant enough to support continuous operations and still accept some losses, and how can dispersed command and control operations be conducted without incurring inefficiencies associated with the dispersion?[14]

One of the most difficult challenges in supporting command decision making is the fusion of data into knowledge. More and more sensors will provide more and more data from more and more locations. A major challenge is converting this information into fused knowledge. What do all the pieces of data mean? Access to more data may actually inhibit, rather than support, better decision making unless this data is fused into reliable knowledge. Different users may need different geographic presentations fused and placed into a common reference grid and may need different levels of detail. Uncertainty regarding the completeness, accuracy, or time of data must be conveyed in its display so that commanders can assess the impact of this uncertainty on decisions. Further, commanders must have the ability and the training to query the "fused" picture to get the understanding they need to carry out their particular piece of the mission. However, the displays all must have a common basis so as to

---

[13]Chairman of the Joint Chiefs. 1996. *Joint Vision 2010*, Joint Chiefs of Staff, Washington, D.C.

[14]For example, the concept of separating selected functions and relying on reach-back to link the elements may have promise but requires further exploration.

convey a common relevant operating picture, enable understanding of command intent, and facilitate self-synchronization.

1.4.3 Support for Deployment of Forces in the Changing Environment

In many cases where the U.S. military will be committed to an actual or emerging situation that destabilizes regional peace or adversely affects U.S. interests in the region, a strategic deployment (from the continental United States) will be required. This need has grown as forward stationing of U.S. forces has diminished, and the early introduction of military capability may become even more crucial. One of the purposes of the military mission of shaping is to facilitate the early approval of overflight, staging, landing, and porting rights at the time of a crisis. With the reduction of forward-stationed U.S. forces worldwide, a significant forward presence may not exist, and U.S. forces would be most vulnerable during their initial arrival, as was the case for the 82nd Airborne's arrival in Saudi Arabia during Desert Shield. A C4I system of systems is needed that can better examine alternative deployments and input requirements, allocate airlift and sealift resources, track deployment movements, and adjust arrival flows. The system of systems must be supported by a global communications network since it must provide the linkage between the home stations of deploying units, the providers of transportation, the supporting forces, enroute movements, the supported forces, and the arrival locations. Obviously, such a system of systems is inherently joint, and often combined, since it must be used by the joint force commander, the military service component commanders, the supporting unified commanders, and the nations providing forces and transportation capabilities. Further, the execution of the deployment must be coordinated with the countries through and into which the flow occurs.

A companion C4I requirement for operating in that environment is the capability to support a reduced logistics footprint, with most of the support needed by U.S. forces provided directly by producer-to-user delivery rather than delivering, receiving, storing, and subsequently redistributing major quantities of materiel in-theater. To meet this requirement, C4I systems need to provide in-transit visibility (not unlike that perfected by Federal Express) and problem detection and movement adjustment capabilities such as that used by much of the trucking industry, and be sufficiently adaptable to support deliveries to small, dispersed, and mobile forces. Again, this system must support the providers (often located in the continental United States), the transportation system, and the eventual recipient of that support, who may be mobile.

A challenge to regional deployments is the missile threat, particularly short- and medium-range ballistic missiles and cruise missiles. While each of the military services may provide some capability for defense against missile attack, it would desirable to rapidly phase in and integrate these capabilities upon initial deployment. Likewise, protecting the arriving forces from air attack will be an important first task involving elements of each of the services. While clearly a critical initial task, an effective air and missile defense must be sustained for both fixed assets and mobile forces. In that environment, C4I and related surveillance and reconnaissance capabilities will need to provide a common air picture, reduce sensor-to-shooter time lines, and integrate service weapon systems into the overall joint mission.

Air power may be the earliest arriving capability and will most likely be a combined effort of contributing nations and elements of the U.S. Air Force and U.S. Navy. Accompanying C4I systems will need to provide the means to determine the most appropriate air assets to allocate to each mission, and disseminate this information in time to allow the missions to be prepared adequately and to be responsive to moving as well as stationary targets. These C4I requirements apply to the Joint Air Operations Center, each service component command, and the air command elements of the contributing nations' air forces.

1.4.4 Proliferation in the Use of the U.S. Military for Sustainment and Support Operations (Military Operations Other Than War)

Current military planning for advanced C4I capabilities is based largely on scenarios in which forces are employed against traditional adversaries in relatively traditional conflict situations. While this focus of planning is generally reasonable, planning must also be sufficiently broad to take into account the likely use of the U.S. military in a much more varied spectrum of military operations. The commitment of U.S. forces to military operations other than war such as peacekeeping, humanitarian assistance, disaster relief, and non-combatant evacuation operations places different demands on C4I systems and may require some different C4I capabilities and/or equipment.

U.S. forces are and will continue to be employed to conduct operations other than war, stability and support operations that cover a wide spectrum of very different missions. Military operations other than war, in contrast to more traditional military operations, can be characterized by (1) forces tailored to accomplish the specific stated mission, which often will involve creating non-standard and non-traditional organizations from elements of other organizations; (2) a need for greater coordination and interoperation with government and non-government agencies; (3) the operation of these tailored forces with new command organizations; (4) forces limited in size;[15] (5) forces that are dispersed and require greater operational independence; (6) restrictive rules of engagement aimed at reducing the potential for undesired escalation, and providing clear limits on the force, and which are understood by potential adversaries; and (7) the potential for undesired escalation or "mission creep" without having the proper force to deal with the new or expanded mission.

For operations other than war, requirements for C4I may entail some of the following issues:

- *Intelligence collection and analysis.* In traditional military operations, the enemy is reasonably well defined; in operations other than war, changing environments and situations may lead to rapid, radical shifts in the definition of the enemy. Intelligence for operations other than war is more focused on individual human beings rather than vehicles or weapons platforms. Thus, intelligence efforts (and hence C4I systems) for operations other than war must have a greater focus on human intelligence—scout patrols, informants, and the like. Operations other than war have a different set of information requirements, such as the need for a great deal of detail on a small area (e.g., the layout and shape of a particular room and the route to that room in a building in which a particular group of people is located). And finally, because in operations other than war forces are often inserted into a situation in which political and historical factors may be highly significant, intelligence analysis must include such contextual factors.

- *Combat Identification and Identification Friend or Foe.* In operations other than war, hostile parties may not identify themselves (e.g., with distinctive personnel uniforms or vehicle insignias). A hostile party may be an individual from the same population that U.S. forces are trying to help, or a large group of refugees on the move that may overwhelm available resources. Furthermore, "hostile" behavior may not even be easy to identify.

- *Planning and coordination.* Because DOD planning tools are for the most part oriented toward major conflict, they often do not provide the

---

[15]Small forces are preferred for operations other than war because they minimize political concerns about undue U.S. involvement, both in the host nation and in the United States. In addition, small forces are much easier and faster to deploy, characteristics that are needed in crisis response situations. Finally, when forces are oversized relative to the job that needs to be done, certain capabilities go unused. The presence of non-useful personnel not only consumes resources; it also leads to boredom and complacency, factors that impede operational readiness and capability.

granularity needed to manage the relatively small forces that are generally deployed for operations other than war. For example, a force sized for such an operation might in its entirety be composed of a couple of battalions-worth of individuals, with platoon- and squad-sized units providing critical functions, whereas planning tools for a major conflict might quantize components by battalion-sized units.

- *Tactical connectivity.* Higher-frequency wireless communications are generally limited to line-of-sight connections. Passing a message from one point to another thus requires either a direct line-of-sight connection or relays that can provide intermediate connection points. When a small force is responsible for a large area (as is the case in distributed expeditionary operations), the density of relay nodes is low, distances between relay nodes are large, and connectivity thus may be more intermittent for patrols communicating with field headquarters. Satellite-based or unmanned aerial vehicle-based communications are an obvious solution, and a number of programs now under way provide such intermediate nodes.[16]

- *Coordination with non-military organizations.* Non-DOD U.S. government agencies, inter-governmental organizations such as the United Nations, indigenous agencies such as the local police force, and non-governmental organizations (and perhaps non-compliant or even belligerent parties) often play key roles in operations other than war, and effective command and control requires communication with them. A high degree of interoperability between U.S. communications equipment and the civilian communications infrastructure, for example, can support non-governmental organizations, thus helping to build trust and good working relationships.

- *Command and control over junior personnel at a distance.* Because of the potential for inadvertent escalation of an interaction between U.S. forces and others (e.g., indigenous civilians or military personnel), troops in the field must often think before they act, whereas a traditional military operation would place a premium on their acting (or reacting) very quickly. Situation assessment must be done in real time by the very junior personnel (privates and corporals) who do the real work in the field. Supporting these junior personnel at a considerable distance can be problematic because many contextual cues are not available to an off-site senior commander. Such field personnel would have greater need for technologies that support consultations and assessment (e.g., laptop computer access to intelligence databases, translation and language services, remote

---

[16] For example there is an effort under way to create a version of the Trojan Spirit system, which provides satellite-based access to intelligence information, that is sized down to be carried by a single vehicle.

conferences with a wide spectrum of possible players) rather than for the capabilities required for combat such as automatic downloads of targeting information.

The C4I implications of military operations other than war and those of counterterrorist operations and operations against the use of weapons of mass destruction will need further study.

1.4.5 Complexities of Exercising Command and Control of Forces in Regional Conflict Environments

Smaller, more capable forces that are widely dispersed will have to depend on firepower from weapons that are not under their direct control. Command and control of these ground forces will be conducted by dispersed and often mobile command elements that also may perform their tasks from multiple locations. To survive and be effective in this environment, dispersed units will need timely, accurate, and common pictures of the combat environment and rapid exchange of target information. Ground line-of-sight communications will not be sufficient, nor will manned or time-consuming relay and switching equipment.

A basic requirement is mobile, agile command and control that can be transferred, for example, from shipboard to ground or from air to ground during the execution of an operation without degrading command and control. The C4I system must be capable of providing robust data, voice, and video communications suitable for collaborative planning. A fundamental and enduring C4I requirement is to facilitate rapid decision making so that the multiple military capabilities of the services can be appropriately integrated and exploited.

The conditions under which U.S. forces are deployed to support military operations other than war may well become more characteristic of some wartime operations in the future. Urban warfare in particular has many of the same characteristics as military operations other than war, e.g., an orientation toward individuals rather than platforms, and a difficulty in separating combatants from non-combatants. In other scenarios, smaller land forces—relying in part on C4I technologies—might be used to control larger expanses of territory, much as forces deployed in operations other than war today do.

1.4.6 Strategic Vulnerability of Infrastructure to Information Attack

The growing dependence of the United States on its national information infrastructure, as well the dependence of other elements of its infrastructure (e.g., electric power, transportation) on information technology, poses potential strategic vulnerabilities that are without precedent. Exploitation of these vulnerabilities by an adversary poses the risk of asymmetric warfare or conflict, in which an adversary does not directly challenge U.S. military might but rather seeks to do damage to the United States in ways that do not require large military forces and where the source of the attack is difficult to identify with certainty.

A further concern is that the U.S. military itself is highly dependent on the U.S. national infrastructure for C4I (information and communications) as well as other services. Thus, a successful attack on the U.S. infrastructure might well have the additional effect of compromising traditional U.S. military readiness and ability to respond militarily.[17]

1.5 EXPECTED INFORMATION TECHNOLOGY TRENDS FOR C4I

Rapid development of information technology and the expectation that C4I technology can dramatically increase force effectiveness have made this technology a critical element of future military modernization. The time constant of progress in information technology, computers, and communications is measured in months, not years. Hardware technologies will continue to evolve at a rapid pace to produce significantly improved capabilities at ever-lower cost—an order-of-magnitude improvement in performance every 5 years for the same cost is likely to continue to be the norm for progress in computing capability (Moore's law), memory and storage capacities, and communications speed. Academic research and the commercial sector are, and will continue to be, the primary sources of fundamental advances in information technologies. Industry exploits these advances, developing and manufacturing high-volume, low-cost, high-reliability products and setting most of the relevant standards. This driving force and dominant market for this expanding capability will continue to be the commercial marketplace, and the same level of basic technology will be readily available to all comers. A key challenge to DOD and the services will continue to be to develop an appropriately responsive acquisition system that can procure, deploy, and exploit these commercial hardware and software capabilities for the military in a timely and cost-effective way.

Much of, although by no means all, the sensor technology essential to C4I systems is specifically developed by the military and for military ap-

---

[17]See President's Commission on Critical Infrastructure Protection. 1997. *Critical Foundations: Protecting America's Infrastructures*, Government Printing Office, Washington, D.C.; Defense Science Board. 1996. *Report of the Defense Science Board Task Force on Information Warfare-Defense (IW-D)*, Office of the Under Secretary of Defense for Acquisition and Technology, Washington, D.C.

plications. The pace of growth in capability is slower than for the base information technologies. Continued focused investment by DOD is expected to maintain a significant margin of leadership in critical sensor technologies.

1.5.1 Computers

The rate of progress predicted by Moore's law means that capabilities seen today in raw processing power of individual computers, as well as associated memory and storage capacities, are about 1% of what will be available at the same cost a decade hence. In addition, major progress will continue on other fronts with significant implications for military application. Decreases in physical size, power consumption, and cost will lead to expanded flexibility and scale of application at the systems level. Expanded and qualitatively more capable applications will become available. For example, more highly automated decision-support systems using intelligent agents will be able to search large databases, including images and other non-coded information, for specific information and features, process the results, and present tactical alternatives to a commander. Continued rapid progress will be made in technologies enabling easier human interaction with computers, including spoken input, high-resolution personal heads-up (e.g., helmet-mounted or windshield) displays, and distributed wearable systems.

1.5.2 Communications

The trend in information distribution and control systems is toward a communications medium that is completely transparent and robust to the military user. These systems will provide global coverage, consisting of highly automated digital networks utilizing both military and commercial transmission media. Current and future developments will enable multimedia service (voice, data, video) to all military users.

Key areas of progress in communications technology applicable to C4I will include advanced video and data compression techniques to transfer expanded information sets through limited-bandwidth channels; wireless wide area network/local area network packet-switched networks utilizing mobile base stations; wider-bandwidth optical communications networks for low-cost, robust terrestrial connectivity; advanced waveforms to maximize coding gain; advanced modulation approaches to increase bandwidth efficiency, given the pressures on military spectrum allocation; "software" radios that provide broadband digital processing; and multifunction, multiband phased array antenna technology that will find application in both communications systems and sensor development.

1.5.3 Sensors

The capability of active and passive multispectral high-resolution sensors in all physical domains (acoustic, thermal, electromagnetic, electro-optical, nuclear, biological, and chemical) is expected to progress at a pace somewhat slower than that of the base information technologies, but still at a rate that will yield impressive opportunities for application to all types of military systems. Continued miniaturization of these sensors and their associated processing units will make them deployable on a variety of platforms, including spacecraft, unmanned aerial vehicles and manned aircraft, land vehicles, ships, and personal battlefield systems. For example, radar technology advances are expected in solid-state transmit/receive modules for higher output power, greater direct current to radio frequency conversion efficiency, increased miniaturization, and wider frequency band operation. Multispectral imaging sensors will prove to be of significant military value in detecting manmade and natural objects.

Technologies for geospatial referencing (such as the Global Positioning System and enhancements to it) that enable the location of targets, events, and friendly forces will also be important. Such technologies confer the ability to register events and objects in the same coordinate system, and underlie the ability to generate a common operating picture.

Some of this capability, originally military in its focus, will become readily available at low cost in the commercial world; some will be specifically developed by the military for its unique requirements. Examples of widely available technologies that were once predominantly military include low-cost Global Positioning System devices and satellite imaging.[18] Examples of military-unique sensor systems include the Airborne Warning and Control System, the Space-based Infrared System, and the Joint Surveillance Target Attack Radar System. DOD will be faced with determining and implementing the appropriate and timely application of this wide array of technologies.

1.5.4 Weapons

Future weapons systems will have integrated digital information subsystems (versus simply having digital communications) that are tightly integrated with the overall C4I system of systems. This capability will allow information available on individual platforms to be simultaneously shared and acted upon across the battlefield (and airspace). Targets acquired by sensors in ground systems and aircraft will be seen concur-

---

[18]Of course, commercial imaging satellites do not provide the resolution that military satellites provide, although commercial image quality will be adequate for many purposes.

rently by multiple platforms and will be rapidly targeted by surface weapons, given pre-established rules. Over time, the value of remote, precision weapons will increase relative to that of other platforms (e.g., tanks, airplanes) as long as the challenge of target identification is solved.

1.6 DOD BUDGET AND ORGANIZATIONAL STRUCTURE FOR C4I

1.6.1 Budget

In a defense budget on the order of $257 billion for FY 1998, spending on C4I is widely quoted as approximately $40 billion,[19] but this figure does not represent an official DOD budget category. The DOD budget category for intelligence and communications is approximately $30.4 billion. But association of this particular figure or any other figure with spending on C4I must come with several caveats and cautions. Among them are the following:

• C4I programs are scattered throughout the 11 primary DOD budget categories.[20] For example, the account for strategic forces includes some funding for C4I systems intended for command and control of the strategic forces. However, these systems can be used to provide connectivity to the general-purpose forces as well. (A good example is Milstar—originally a communications satellite for strategic use, it is now used for non-nuclear purposes as well.)

• C4I programs per se are distinct from C4I systems embedded within weapons systems. For example, neither the radar for an F-22 fighter nor the radar for a Patriot air defense system would be counted as C4I programs, though they are clearly C4I systems.

• C4I programs include systems for intelligence work, much of which is "black" and thus not known publicly.

• Programs for surveillance and reconnaissance are not always included in an accounting of C4I systems.

One public estimate of the amount of "electronic content" in the overall defense budget provided by the Electronic Industries Alliance is ap-

---

[19]General Accounting Office. 1998. *Defense Information Superiority: Progress Made, But Significant Challenges Remain*, GAO/AIMD-98-257, General Accounting Office, Washington, D.C.

[20]These categories are strategic forces; general-purpose forces; intelligence and communications; air and sealift; guard and reserve forces; research and development; central supply and maintenance; training, medical, and other; administrative and associated costs; support to other nations; and special operations.

proximately $51.5 billion in FY 1998, a figure that includes acquisition as well as operations and maintenance.[21] This figure covers all possible categories of C4I systems, including those for surveillance and reconnaissance.

These data are provided to give the reader a sense of scale of C4I within the defense budget. But it should be noted from the outset that because the committee does not seek to provide detailed programmatic guidance, the analysis, findings, and recommendations of this report are essentially independent of the numbers discussed above.

1.6.2 DOD Organizational Structure for C4I

Responsibility for the development, procurement, operations, and maintenance of specific C4I systems generally lies with the services. However, CINCs[22] and field units do have some discretionary budget authority to purchase systems below a certain cost threshold. The Defense Information Systems Agency has the primary responsibility for maintaining defense-wide C4I infrastructure (e.g., that for long-haul communications). Research and development into information technologies that may eventually be integrated into actual C4I systems is undertaken by the Defense Advanced Research Projects Agency and the various service research arms. The National Security Agency plays a key role in providing technologies and products for information security.

Oversight of C4I system acquisition is performed by a myriad of organizations and offices. Some of the most important are the Under Secretary of Defense for Acquisition and Technology, the ultimate authority within DOD on acquisition matters; the Assistant Secretary of Defense for C3I, the focal point of DOD policy with respect to matters related to C4I and information superiority (and also today the DOD's Chief Information Officer); the Joint Requirements Oversight Council, an organization that validates requirements and military needs for "major" C4I systems; the Defense Acquisition Board, which is chaired by the Under Secretary of Defense for Acquisition and Technology and advises on individual acquisition programs and generally on acquisition policies and procedures; the

---

[21]Electronics Industry Association press release, "EIA Ten-Year Forecast Projects 14% Growth in Electronics; Defense Market Remains Stable," October 8, 1997; available online at <http://www.eia.org/pad/press/files/9710/97-59.htm>.

[22]CINC, an acronym for "commander-in-chief," refers to the commander of a specified or unified combatant command. The term "CINCs" refers to the commanders of the combatant commands. The combatant commands include the U.S. European Command, U.S. Pacific Command, U.S. Atlantic Command, U.S. Southern Command, U.S. Central Command, U.S. Space Command, U.S. Special Operations Command, U.S. Transportation Command, and U.S. Strategic Command.

Major Automated Information Systems Review Council, which is mandated to advise the Assistant Secretary of Defense for C3I on decisions regarding major individual automated information system acquisition programs; and the Directorate for C4 Systems of the Joint Staff, which has responsibility for command, control, communications, and computer (C4) systems, especially with respect to interoperability and integration.

This listing of organizations is far from complete—indeed, the committee was struck by both the multiplicity of organizations and offices with some responsibility for C4I matters, and the relative rapidity with which the organizational structure for C4I has been evolving.

1.7 CHALLENGES TO THE EXPLOITATION OF THE MILITARY LEVERAGE OF C4I

While the complexities and uncertainties of the future produce a major set of challenges to the development, integration, and fielding of the "right" set of C4I systems and processes, the U.S. military faces another set of challenges in implementation. These challenges are of both a technical and management nature, and most are specific to the military system. They are challenges that can be, and indeed are being, addressed now. The remainder of this report is devoted to the committee's view of the nature of these challenges, the state of progress in addressing them, and the actions that must be taken to deal with them more forcefully and effectively. This report addresses challenges in three areas: (1) achieving interoperability, (2) ensuring security and systems availability, and (3) evolving the military culture and business processes to enable what is required in tomorrow's world.

First, C4I systems must be interoperable so as to support joint and combined operations and the necessary interaction with government and non-governmental organizations in an environment in which the sophistication of C4I systems available to various units (or coalition partners) will surely span a spectrum of capability. Achieving this level of interoperability poses technical as well as cultural and process challenges. Significant technical dimensions include design tensions between immediate and future needs; tensions between applications-specific needs and the needs of the entire system of systems; inability to anticipate all relevant scenarios for use, resulting in an inability to anticipate which systems need to interoperate; extent of backward compatibility to be designed into systems; difficulties of anticipating a sustainable technology environment; inherent difficulties of system integration; and synchronization of interdependent programs. A number of cultural dimensions also affect efforts to achieve C4I interoperability, including the profound differences between peacetime and wartime missions, rapid management turnover that is characteristic of most government organizations, use of service-based acquisition, doctrine for interoperating with heterogeneously equipped forces, a lack of resources to pursue C4I integration as a high-priority budget item, line-item budget accountability, and the need to operate in coalitions that are quickly assembled and cannot be anticipated.

Second, C4I systems must be secured against information attacks. With increased reliance on C4I systems as well as an increased use of commercial technologies to build these systems comes a new and increased set of risks associated with the vulnerability of these systems to attack. Here, too, there are technical and cultural dimensions. Technical dimensions include the need for good automated tools for checking and inspecting network and system configurations and tools that allow the rapid and high-confidence identification of a cyber-attacker and retaliation against such attackers. A distinction must be maintained between the attacker whose intent is to disrupt or corrupt the C4I system and one whose intent is to monitor and collect information from one. Cultural dimensions include the need to promulgate a defense-wide awareness of information security (ranging from accountability to providing good information security support) and a legal constraint and military tradition of refraining from involvement in domestic security affairs.

Third, the base technologies of C4I evolve at such a rapid rate that cultural and technical challenges arise with respect to how, when, and what aspect of the technology can best be exploited to significantly increase the leverage of information systems in military operations. Infusion of technical skills in the military workplace will be required along with bringing doctrine abreast of the advances in technology. Also, leadership skills will need to be honed to take account of the technical and doctrinal shifts brought about by the potential inherent in advanced information technology. Indeed, the very fact of revolutionary changes in military operations brought about by advanced C4I systems poses enormous leadership challenges for the U.S. military, which as an institution practices well-justified conservatism. Finally, it is important to highlight the challenge to the whole acquisition process, which must take into account the rapid pace of change in information technology and the dominant role of the commercial sector in driving technological advances. The challenge is exploiting the rapid advances in information technology at a time when many, if not most, of these technologies are available through the commercial market with an acquisition system not designed to exploit rapid acquisition. Each of these three challenges, then, is discussed in the following chapters.

2

Interoperability

2.1 WHAT IS INTEROPERABILITY AND WHY IS IT IMPORTANT?

The full realization of Joint Vision 2010[1] and the revolution in military affairs, discussed in Chapter 1, is predicated on a concept of information superiority enabled and supported by a network of C4I systems—one whose constituent elements interoperate and cooperate to support the entire warfighting hierarchy, in the context of joint and coalition operations. Interoperability of C4I systems is a key enabler of the overarching operational goal of force integration—the fusing of the services and coalition partners into a unified military force that achieves high military effectiveness, exploiting and coordinating the individual force capabilities. Achievement of a high level of interoperability requires a commensurate level of effort and resource prioritization throughout DOD. Today, DOD is just at the beginning of refining and even establishing the processes and organizations to respond to future needs for C4I interoperability.

2.1.1 What Is Interoperability?

Interoperability is a broad and complex subject rather than a binary attribute of systems. C4I interoperability is a key enabler for the conduct

---

[1] Chairman of the Joint Chiefs of Staff. 1996. *Joint Vision 2010,* Joint Chiefs of Staff, Washington, D.C.

of effective, collaborative, multi-service military operations across a wide spectrum of scenarios, and successful conduct of operations is the ultimate test of whether an adequate degree of interoperability is being achieved. Joint Chiefs of Staff Publication 1-02 defines interoperability at both the technical and operational level (Box 2.1).[2] Operational interoperability addresses support to military operations and, as such, goes beyond systems to include people and procedures, interacting on an end-to-end basis. Implementation of operational interoperability implies not only the traditional approach of using standards but also enabling and assuring activities such as testing and certification, configuration and version management, and training. These definitions of operational interoperability encompass the full spectrum of military operations, including intra-service/agency, joint (inter-service/agency), and ad hoc and formal multinational alliances.

Interoperability at the technical level (see Box 2.1) is essential to achieving operational interoperability. An issue that arises between systems rather than between organizations, technical interoperability must be considered in a variety of contexts and scopes, even for a single mission. Consider the theater missile defense mission, which is likely to require that data be:

- *Exchanged among elements of a weapon system.* For example, the Patriot air defense system uses a defined message format and data link to exchange information within batteries and between batteries to share target information and coordinate defensive actions.
- *Exchanged between weapons systems of a single organization or service.* For example, the Theater High Altitude Area Defense system (under development) will provide theater ballistic missile tracks to Patriot systems.
- *Exchanged between weapons systems of different services.* For example, a Navy AEGIS radar may report tracks to an Army Patriot radar.
- *Shared and "pooled" at the joint task force command and control systems level (or higher) in order to achieve synergy and added value.* For example, Patriot, AEGIS, and Airborne Warning and Control System data may be combined to develop a common operating picture and to control and coordinate all the systems sharing data.

The range of complexity of requirements for data flow in such a mission underscores the significance of interoperability at every level.

---

[2]Joint Chiefs of Staff. 1998. *Department of Defense Dictionary of Military and Associated Terms*, as amended through December 7, 1998, Joint Publication 1-02, Joint Chiefs of Staff, Washington, D.C.

> **BOX 2.1 Interoperability Defined**
>
> *Operational Interoperability*
>
> The ability of systems, units, or forces to provide services to and accept services from other systems, units, or forces and to use the services so exchanged to enable them to operate effectively together.
> —Definition (1) in Joint Chiefs of Staff, *Department of Defense Dictionary of Military and Associated Terms*, as amended through December 7, 1998 (Joint Publication 1-02) and Chairman of the Joint Chiefs of Staff, *Instruction 6212.01A: Compatibility, Interoperability, and Integration of Command, Control, Communications, Computers, and Intelligence Systems*, June 1995.
>
> The ability of systems, units, or forces to provide services to or access services from other systems, units, or forces, and use the services to operate effectively together.
> —DOD Directive 5000.1, "Defense Acquisition," March 15, 1996.
>
> *Technical Interoperability*
>
> The condition achieved among communications-electronics systems or items of communications-electronics equipment when information or services can be exchanged directly and satisfactorily between them and/or their users. The degree of interoperability should be defined when referring to specific cases.
> —Definition (2) in Joint Chiefs of Staff, *Department of Defense Dictionary of Military and Associated Terms*, as amended through December 7, 1998 (Joint Publication 1-02).
>
> Interoperability is the ability of systems to provide dynamic interactive information and data exchange among C4I nodes for planning, coordination, integration, and execution of Theater Air Missile Defense operations.
> —Joint Theater Air Missile Defense Organization (JTAMDO). 1997. *JTAMDO Master Plan*. JTAMDO, Joint Staff, Department of Defense, Washington, D.C., Chapter 7.

One source of interoperability problems is incompatibilities in independently selected versions (e.g., software releases) of the same system. Thus, if one unit has standardized on version A of a given system and another on version B, capabilities supported by one system and not the other may well interfere with seamless interoperation between the two units. The committee observed several such situations in its visits to exercises. Also, just as differences in modes of operation across the services can lead to non-interoperability, so can organizational differences within a service also lead to intra-service incompatibilities. One example that the committee heard about involved information security procedures and practices that were different in the U.S. Atlantic and U.S. Pacific commands, presenting difficulties for units that are reassigned from one theater to another.

When thinking about C4I it is important to understand the distinction between joint systems and systems that are interoperable (Box 2.2). A system is designated as joint either to support an efficient buying decision for two or more services that will use it, or because the system will be subject to joint command. By contrast, to meet requirements for interoperability, services' systems must be able to share data in a timely, reliable manner that is operationally useful, and must operate across service or agency boundaries to support joint missions. Interoperability does not necessarily imply joint (multiple service) programs; interoperability can and must be achieved without jointness. Joint programs are but one of a number of management approaches for achieving interoperability of systems among the military services.

Although many view interoperability as an issue arising in the context of two or more services (or nations), fielding a wide variety of mature systems built with little attention to supporting joint or coalition operations,[3] in fact its sphere is broader. Indeed, during its site visits the committee heard several examples of C4I systems owned and operated by the same service that have difficulties in interoperating. For example, in one visit, the committee observed that with the Army Forward Area Air Defense Command, Control and Intelligence System and the Maneuver Control System there were difficulties in overlaying data from one system with data from the other.

Finally, although it is a critical enabler for military operations, interoperability must be recognized as just one of several technical attributes of any system of systems. Indeed, other attributes will sometimes be in competition with interoperability and with each other; an appropriate balance must be sought. For example, there are trade-offs between security and interoperability. Interoperability can promote an attacker's access to diverse systems, thus facilitating the rapid spread of attacks. Also, ad hoc work-arounds to overcome a lack of inherent interoperability can

---

[3]For example, in the Gulf War, C4I system incompatibilities made it impossible to electronically transmit the Air Tasking Order to Navy carriers, making delivery of paper copies necessary.

BOX 2.2 Interoperable Systems Are Not the Same as Joint Systems

- An airborne command and control system for air-to-air engagement could be designed as an Air Force system but its air picture would have to be displayable by Army air defense units and Navy ships. Likewise, the Navy might design a system for AEGIS air defense of the fleet, and the Army might design a surface air defense system. If an integrated air picture is to be shared by all three services, the data from each of the three service air defense command and control systems, which each provide an air picture based on data from the system's own sensors, must be exchangeable. Such service air defense command and control systems would be interoperable but not joint.
- The Army might design a surface-to-surface artillery command and control system. The Navy might design a ship-to-shore naval gunfire system. Because targets might be able to be attacked by either the Army or Navy weapons, it would be useful to be able to pass attack orders between the two systems—setting a requirement for interoperability but not jointness. As the Navy develops its Cooperative Engagement System to integrate both naval surface-to-air and surface-to-surface firepower, it would be useful to be able to share its target information with Army air defense and field artillery systems, and likewise share the Army system capabilities with the Navy system. If, however, it were ever desired to receive requests, determine targets, and automatically to fire on air and surface targets using either Army or Navy air defense or surface-to-surface firepower, it would make sense to develop a joint fire control system. Such a system would be considered joint because it would be employed by both services and would control the use of resources from both services.
- The U.S. Transportation Command could develop a command and control system to allocate air and sealift. The consumers of the command's transportation services may never need to use the system directly themselves, but would be interested in interoperability considerations such as the ease (form) of inputting lift requirements and of reading the output to determine what items need to be prepared for each lift asset. If this same system were to allow consumers to themselves conduct assessments of alternative deployments, it would then be considered a joint system because input form, run time, and output form might be of critical concern to consumers from all of the services. However, this does not mean that such a system would have to be developed and procured by a joint program office as opposed to that of a single service lead (with requirements input by the other services).
- With the advent of the Joint Force Air Component Commander, who integrates and can employ all air assets, the next version of software to

continues

> produce a joint air tasking order should probably be a joint development. This system should be used by both the Air Force and the Navy, should accommodate the uniqueness of each service, should decide or assist in the decision of the best assets to be employed (regardless of service), should integrate service assets (e.g., naval attack aircraft and Air Force tankers), and should provide a single, integrated order. This joint air tasking order development system could be used by both services to produce their air tasking orders when each is operating separately and could be used to produce a joint air tasking order when a Joint Force Air Component Commander has been established.

introduce many hard-to-manage security problems. Another trade-off is the potential for interoperability problems posed by the introduction of new security features into part of a larger system of systems. Thus, in thinking about overall system functionality or performance, security requirements such as confidentiality, authentication, non-repudiation, integrity, and system availability must be considered together with interoperability.

2.1.2 Why Interoperability Is Important

Why is interoperability, which is so difficult to achieve, so essential to implementing current doctrine as well as emerging concepts of operation?

Experience in operations such as Desert Storm and Bosnia, as well as evidence from recent experiments and exercises, points to the dramatic improvements in operational effectiveness that are achievable using highly capable C4I systems. The leverage provided by a common operating picture and the rapid decision-making ability associated with it can dramatically change the pace, nature, and geographic range of engagement, providing major advantage to forces so enabled. Interoperability is a key to realizing these advantages.

Interoperability is also an important factor in operational efficiency. Where interoperability is lacking, there is the likelihood that multiple systems are performing the same functions, or that information is being manually entered or processed multiple times. And lack of interoperability also means that personnel have to resort to work-arounds. Where interoperability is not in place, necessary transfer of information between systems may require speaking over a voice link or rekeying data from printouts or handwritten notes, processes that are not only inefficient but error-prone as well.

Military operations are typically joint, requiring that the C4I systems of multiple services work together effectively. Both the generally unpredictable nature of military contingencies and the wide range of non-traditional operations mean that forces and weapons are likely to be combined in novel, unanticipated ways to meet operational requirements and that their C4I systems may need to interoperate in ways not explicitly planned for in advance. Also, the new operational emphasis on rapid force projection, and the concept of early entry to halt an invasion, mean that there will likely be less time during a deployment to fix interoperability problems. Finally, the increasing size of the area over which combat operations take place—and thus the number of possible forces and weapons that must coordinate their attack—means that data is increasingly being exchanged between sensors, weapons, and systems that previously operated in a stand-alone manner. To meet such operational requirements, the different elements of the C4I system of systems will need to be more interoperable.

Many important military missions require a high degree of interoperability to support cross-service collaboration. Some specific instances in the area of joint operations include the following:

- Close air support, which requires that Army ground troops be able to communicate their air support needs to Air Force ground attack airplanes in a timely and accurate fashion;
- Suppression of enemy air defenses, which in general requires the coordinated use of missiles and aircraft operated by multiple services;
- Theater missile defense, where, as noted above, data may be shared between weapons systems of different services or shared at the joint task force command and control systems level (or higher);
- Regional air defense, which requires the coordination of many air defense assets, from missile batteries and radar on the ground to airborne surveillance platforms and air defense fighters; and
- Deep-strike attacks and interdiction of enemy forces behind the front lines, which both require the coordinated use of airspace, strike aircraft, ground- and sea-based missiles, and long-range artillery.

There is also ample evidence from experience that inadequate interoperability can cause major problems and significantly reduce military effectiveness. A recent report by the Secretary of Defense noted that "from Grenada in 1983 to Operation Desert Storm in 1991, joint operations have been hindered by the inability of forces to share critical information at the rate and at the locations demanded by modern warfare."[4] Recent examples include the following:[5]

- During Operation Urgent Fury, the invasion of Grenada, the Marines and the Army Rangers could communicate only through offshore relay stations, because their use of radio frequencies was uncoordinated. As a result, the Marines did not know in one instance that the Army Rangers were "pinned down without adequate armor."[6]
- During the Joint Warrior Interoperability Demonstration in 1996, problems associated with network configuration did not support "timely interoperability" with coalition forces. According to the Joint Warrior Interoperability Demonstration report, limitations of the multilevel security systems, which were intended to allow information to be delivered to coalition forces, required manual intervention even for use of simple applications such as e-mail. This need for manual intervention "made it extremely difficult to conduct U.S./coalition collaborative planning since information . . . was never fully synchronized for both U.S. and Allied planning requirements."[7]
- According to the General Accounting Office, 43 "significant interoperability problems" associated with 15 C4I systems and weapons, such as the AEGIS and Patriot systems, were identified by the Joint Interoperability Test Command during four joint exercises held in 1996 and 1997. These interoperability problems, most of which "were caused by

---

[4]William S. Cohen. 1998. *Secretary of Defense Report to Congress: Actions to Accelerate the Movement to the New Workforce Vision*, Department of Defense, Washington, D.C. This report also tasked a study to develop an improved, cross-service process for developing joint capabilities.

[5]The Joint Interoperability Test Center of the Defense Information Systems Agency is responsible for testing and evaluating C4I acquisitions and systems, as well as identifying and solving C4I interoperability problems. As part of its work, it compiles the quarterly compilation of lessons learned, which addresses "C4I interoperability problems/issues related to Joint/Combined C4I and integration of information systems within the Defense Information Infrastructure." For additional information, see the Joint Interoperability Test Center home page at <http://www.jitc.fhu.disa.mil>.

[6]Col. Stephen E. Anno and Lt. Col. William E. Einsphar (no date). *Command and Control and Communications Lessons Learned: Iranian Rescue, Falklands Conflict, Grenada Invasion, Libya Raid*, Air War College Research Report No. AU-AWC-88-043, Air University, United States Air Force, Maxwell Air Force Base. Information can be obtained through the Joint Electronic Library at <http://www.dtic.mil/doctrine/jel/research_pubs.htm>.

[7]Defense Information Systems Agency (DISA). 1996. *Joint Warrior Interoperability Demonstration 1996 Report*, Defense Information Systems Agency C4I Modeling, Simulation and Assessment Division, DISA, Arlington, Va.

system-specific software problems," could potentially "result in the loss of life, equipment, or supplies."[8]

In short, interoperability is essential to operability—that is, forces cannot operate effectively without a high degree of interoperability among their systems. Unfortunately, interoperability is often treated as a potentially desirable but nonessential, element of C4I programs, and a sufficient degree of interoperability, especially inter-service, is not currently seen by managers as a pass-fail criterion for their programs. Consequently, interoperability requirements tend to be one of the first things sacrificed when budgets force program cost reductions.

That said, universal interoperability is neither achievable nor necessary. Not every C4I system on the battlefield needs to interoperate with every other one. Nor is universal interoperability—which might be thought of as allowing all information in all systems to be seamlessly exchanged and interpreted—technically feasible, given the rate of change in both technologies and missions. The importance of achieving interoperability, determination of what and how much is sufficient, and decision making about allocation of resources to achieve interoperability can be addressed only in an operational context.

2.1.3 Dimensions of Technical Interoperability

On a digital battlefield, sensors generate bits, communications channels transmit bits, computers process bits, commanders act on information represented as bits, and weapons are directed by messages composed of bits. These bits are the underlying electronic representation of data and information, and to be used they must be interpretable according to some agreed-upon definitions. For two C4I systems to effectively interoperate, they must be able not only to exchange relevant bitstreams but also to interpret the bits they exchange according to consistent definitions—merely providing information in digital form does not necessarily mean that it can be readily shared between C4I systems.

Interoperability also requires that systems are interoperable at the data level—that the format and semantics of the data are also coordinated so as to permit interoperation. One significant instance where this requirement arises is in the exchange of geographical coordinates. To launch

---

[8]General Accounting Office. 1998. *Joint Military Operations: Weakness in DOD's Process for Certifying C4I Systems' Interoperability,* GAO/NSIAD-98-73, General Accounting Office, Washington, D.C.

a missile against a target, it is necessary to know the location of the missile launcher as well as that of the target. The specification of a location implies the existence of a common coordinate system (and hence a model of Earth) within which both target and launcher locations can be specified. Obvious difficulties can arise if the locations of the target and launcher are specified with respect to different Earth models. If the sensor and launcher are not using the same Earth model, a transformation of the sensor-reported location of the target into the launcher's coordinate system will be necessary. Since it has only been relatively recently that the idea of using non-co-located sensors for fire control has become practical, the implicit assumption of identical Earth models for target and launcher may well not be a valid one. Section 2.3.4 discusses some approaches to the data interoperability challenges.

Thus technical interoperability places detailed demands at multiple levels, which range from physical interconnection to correct interpretation by applications of data that is provided by other applications.

2.2 WHY ACHIEVING INTEROPERABILITY IS DIFFICULT

2.2.1 Challenges Common to All Large Enterprises

Experience in the private sector suggests that the following factors (among others) often operate to inhibit or slow achievement of desired system interoperability.

Tension Between Immediate and Future Needs

Operational units (in the DOD context, the CINCs as the warfighting authorities) in an organization often have a perspective very different from that of the planning units (in the DOD context, e.g., the Office of the Secretary of Defense, Joint Chiefs of Staff, and the service chiefs as the policy makers, allocators of resources, and providers). Operational units are concerned with the capabilities of today's systems in the short term, whereas planning units are concerned with the capabilities of tomorrow's systems, over the longer term.

- For the planner, interoperability is a capability that must be designed into a system. For the operator, interoperability is often achieved by working around problems, e.g., deciding what parts of a system to use or not use, creating patches, and modifying policy or doctrine associated with its use.
- For the planner, changes in system capability (i.e., changes in feature and function) are important. To the operator, changes in operating

capability (perhaps enabled by changes in deployed system capability) have greater significance.
- For the planner, operational doctrine and tactics are driven by what can be imagined when the force is fully equipped and the new technology or system is deployed. For the operator, they are driven by deployment (perhaps partial) of a *system* and the resulting capabilities of the *unit*.

Large organizations recognize that operational considerations (e.g., training, doctrine) must be an integral part of system acquisition. But maintaining such a focus is difficult when operators believe (with some justification) that planners are not rapidly responsive to their immediate needs, and the planners believe (with some justification) that an overemphasis on immediate needs will not enable operating units to fully realize the benefits of new capabilities.

Tension Between Local and Global Needs

Optimizing overall system performance requires a full understanding of the trade-offs entailed by different choices. However, individual units within an organization, especially those that seldom interact with other units, are strongly motivated to solve their own pressing problems, even if doing so makes it harder for them to interact with other units. In addition, the fact that many acquisition programs have very long time lines increases the pressure to deploy independently developed solutions. The most likely result of such independent development is a patchwork of systems that are even less interoperable.[9]

Inability to Anticipate All Relevant Scenarios for Use

Many of the most common applications of information technology today were unanticipated when the technology was initially deployed. For example, when the ARPANET (the forerunner of the Internet) was first deployed, e-mail was considered a secondary application, whereas e-mail is today one of the most often used Internet applications. In general, it is difficult to anticipate in detail how information systems will be used— a difficulty that is multiplied in an uncertain environment. For example, a

---

[9]This phenomenon was seen in commercial industry 20 to 30 years ago. Early application of computing technology that automated the functions of individual business units took place in parallel without attention to enterprise-wide concerns, compounding later interoperability problems. Later developments, such as a shift in technology and approach from reliance on departmental minicomputers to use of client-server configurations, helped resolve these problems.

task force that responds to an operational contingency will usually include units that are drawn from multiple services and that have not trained or fought together in the past. Achieving flexibility in such a situation depends heavily on building in a sufficient degree of interoperability. On the other hand, interoperability does not come free. Explicitly adding pairwise interoperability between specific systems that need to work together may be in some cases cheaper or faster than building every system according to a high-level design for interoperability. And there are also possible performance trade-offs; the high-level design may not be locally optimal. Given these trade-offs, it is a large challenge to define the minimum number of instances where interoperability is imperative, and to estimate the incremental value of increased interoperability.

Development of Component Systems by Different Organizations

Different parts of a system of systems are likely to be built by separate parts of the overall organization. For example, parallel efforts are employed to reduce overall development time. Since each organization has a tendency to optimize the solution to its part of the problem, a certain degree of stovepiping—the building of systems that support only some parts of an enterprise and fail to integrate across enterprise units—is the likely result even when all parties have the best of intentions.

Inclusion of Legacy Systems

Legacy systems are in-place systems that are relatively old and were not designed to be easily integrated with current and future information systems, but which remain absolutely essential to the functioning of the organization. Furthermore, they often represent significant investment, so that replacing them with new, more interoperable systems is not a near-term option.

Managing in the Face of Rapid Technological Change

Because the underlying information technologies will certainly improve significantly over the lifetime of a system's development and deployment, it is desirable to plan for the incorporation of these improvements during the later phases of system deployment. Systems designers must thus pay particular attention to two areas:

- *Sustainability of the technological environment selected.* Technology selection and migration strategies have significant implications for interoperability. Initial choices of technologies from which systems will be built have long-lasting consequences, because they in effect freeze an enterprise's infrastructure. But because the information technology industry is so dynamic, even broadly accepted technologies may later be abandoned by the marketplace. As a result, for example, a company that 10 years ago had selected CP/M as its basic operating system would have had to convert long ago in order to remain current. Because maintaining interoperability with systems based on an abandoned infrastructure is in the long run a very expensive task, developers forced by circumstance to develop applications in an obsolete (or soon-to-be-unsupported) technological environment must also have migration strategies to port their applications to a more sustainable environment. On the other hand, the very latest technologies are not always the best choice; to maximize the likelihood of maintaining interoperability, it is prudent to select relatively stable technologies that are achieving widespread adoption and are likely to enjoy longer-term support.

- *Backward compatibility.* To at least minimally protect users' investments in design, applications, and training, and provide at least a limited measure of interoperability across versions, commercial information technologies usually incorporate considerable backward compatibility from generation N to generation $N + 1$, and usually provide tools to facilitate user transition to the newer generation. However, support for backward compatibility is not unlimited, and at some point, support for the earliest generations is usually abandoned. (Thus, generation $N - 3$ may no longer be fully supported.) Indeed, given the rate of evolution of the processing and storage capabilities of the underlying commercial technologies—and the advances in applications that these improvements enable—it is unrealistic to maintain backward compatibility forever. Management must provide guidance to system designers for how long backward compatibility is to be maintained and indicate a strategy for defining, batching, and sequencing system upgrades. In general, configuration control is required to provide operationally required interoperability and minimize deployment and training costs. The problem is made more difficult when the rate at which enterprise-wide upgrades take place is much slower than the rate of progress in the underlying technology (for further discussion in the DOD context, see section 2.2.2).

Heterogeneously Equipped Organizations or Units

Large organizations usually stage the rollout of new generations of technology or applications over a period of time, either to shake out problems before a full deployment or because limitations on budget or other resources such as training do not allow the deployment of a new generation all at once. In all of these instances, the systems deployed must have interfaces that allow some degree of backward compatibility so that they will be able to exchange data. In addition, policy or doctrine must deal with the need for units equipped with new generations of technology to interact with others not comparably equipped (the issue of how to address this question in military doctrine is addressed in Chapter 4).

Proprietary Technologies

Use of off-the-shelf products and subsystems built to commercial standards can reduce costs and development time and can make interoperability easier to achieve. (Box 2.3 describes how commercial standards are established and the basis for their staying power.) A good program definition will have clear criteria for when the use of commercial off-the-shelf (COTS) products is appropriate. Because contractors may be motivated to avoid COTS products to maintain account control and differentiate themselves in the marketplace, incentives are required to ensure the use of COTS products when appropriate.

Inadequacies of Existing Commercial Standards

Of equal importance to the appropriate use of COTS technology is the establishment of clear criteria for when its use is not appropriate, or at least for how difficulties arising through its use will be addressed. Such criteria are required to address security problems that arise in many uses of commercial products such as off-the-shelf operating systems. Another example of the inadequacies of commercial standards is the problem of using Open Shortest Path First routing in an environment where the "shortest path" routing algorithm may yield undesirable results, such as when mobile, low-capacity routers become the shortest path in a battlefield.[10]

Controlling System Requirement Creep

It is the normal behavior of vendors to try to offer some unique perceived benefit over their competitors. If these benefits take the form of

---

[10]Open Shortest Path First routing calls for traffic to be routed through open paths that are physically shortest. In a static environment, path lengths are known, and congestion on any particular path is a function only of the traffic being carried in the network. But in a dynamic environment (e.g., a battlefield), path lengths change unpredictably. In particular, an airborne command post may suddenly find itself in a position that routing through the command post is the routing of choice for Open Shortest Path First. If this happens, all battlefield traffic may be routed through it, thus overloading the node.

BOX 2.3 The Development of Commercial Standards

Commercial standards are set in two primary ways. One method relies on the development of a consensus among private firms, technical experts, customers, and other interested parties. For example, the Internet Engineering Task Force helps to develop the consensus on many of the standards for the Internet. A number of consensus-based standards-setting bodies follow consensus standards development procedures promulgated by the American National Standards Institute. These procedures include open participation of volunteer technical experts in standards-writing committees; consensus agreement among committee members in support of any proposed standard; and elements of administrative due process, such as opportunities for comment and voting by affected parties.

A second approach to standards development in the private sector relies on marketplace competition. When one firm's product achieves a high degree of dominance in the marketplace, its specifications become a de facto (or industry) standard. A variant on the creation of industry standards is their promotion through industry consortia. Examples of such consortia include the Object Management Group and the World Wide Web Consortium. De facto standards largely characterize the world of information technology and communications networks. Indeed, networked computers and communication devices are of little value if they are based on a standard that few others use—there is no one to communicate with. Thus, increases in the number of devices conforming to a standard lead to greater pressure for other devices to conform to that standard.

Both approaches have in common the fact that they are—to varying degrees—supported by either the marketplace and/or a broad base of vendors and customers. This base of support helps to ensure that products incorporating commercial standards will continue to be built and purchased, thus reducing the chance that products will be "orphaned."

None of the above discussion should be taken to imply that commercial standards have no downside. Standards may freeze technology prematurely. User commitments to the use of a prematurely established standard and a hard-to-change infrastructure can then restrict the development and deployment of new and more useful technologies. Moreover, a standard that is popular in the marketplace may not necessarily be the most appropriate for all end-user applications. Nevertheless, it is safe to say that standards have on balance facilitated rather than retarded the growth of the information technology industry and the rapid pace of technological development.

SOURCE: Adapted from National Research Council. 1995. *Standards, Conformity Assessment, and Trade*, National Academy Press, Washington, D.C.

additional system features, components incorporating those features may create additional complications for interoperability. Users must not allow suppliers to drive the system requirements so that they can differentiate themselves. Suppliers must be convinced that they have more to gain by conformance to a common architecture than they have to gain by product differentiation.

Difficulties of System Integration in Complex and Critical Deployed Systems

Systems integration in the DOD environment poses particular difficulties in at least two instances. The first is the need to integrate a large number of components. Since both the components and their specifications are usually incomplete, systems integrators have to discover what combinations of components (down to the specific version or release number) can interoperate successfully. The number of such rules grows much faster than the number of components. A second difficulty is integrating into an operational system that cannot be taken offline. In the absence of systems that replicate essential elements of the operational system, all integration testing must be performed on the operational system, which places many constraints on the changes that can be made. Even in the civilian world, an industrial laboratory is sometimes used to reproduce bugs that are exhibited in a particular customer configuration. One cannot take over the customer's mission-critical systems to do testing; it has to be reproduced in a separate test facility. Similarly, one cannot stop a war to debug the C4I systems. Resolving faults that occur in an engagement often requires offline resources for debugging.

Synchronization of Interdependent Programs

In many large deployments, a number of independently developed components must be brought together to work as a whole. If two systems are to be pairwise-interoperable, design decisions in one program may have an effect on the other program. If the first program is significantly delayed, the other program may have to proceed without those decisions being made, with the likely result that interoperability in the end may be adversely affected. The alternative is delaying the second program, a highly undesirable outcome. Thus, the time lines for developing these components must be synchronized if interoperability is to be effected in a timely manner. Such synchronization refers not only to product delivery at the proper times, but also to matters such as timely decision making and testing within each program so that decisions are made that facilitate interoperability and problems uncovered soon enough to not cause additional difficulties later.

2.2.2 Special Challenges Faced by the Department of Defense

All of the factors described in section 2.2.1 are challenges faced by the DOD. In addition, the unique mission of the DOD poses challenges beyond those typically found in the civilian sector, including the following.

Unanticipated Usage

As noted above, it can be safely predicted that over the lifetime of various C4I systems now and yet to be fielded, some operational scenarios will call for the use of these systems in ways that cannot be anticipated today. Requirements for interoperability of C4I systems, as well as the C4I embedded in weapons and sensors, cannot always be fully anticipated in advance.

Many weapon and sensor systems were designed to operate in relatively loose coordination with other systems. Today and in the future, these weapon and sensor systems are intended, and will be expected, to operate much more cooperatively, thus placing more stringent demands on C4I systems. Moreover, the flip side is also true: as new C4I capabilities become available, a weapon or sensor system should be able to exploit them.

Unforeseen C4I linkages also arise when old weapons systems are given new missions. For example, a nuclear delivery system such as the B-1 bomber may be assigned to conventional bombing missions with substantially different C4I requirements. In this case an interoperability problem is created at the physical layer (Box 2.4); the B-1, originally designed for strategic nuclear missions, cannot receive digital downloads of information that would enable it to retarget weapons while in flight. A second example is that the Patriot missile system, originally deployed for defense against aircraft, is now used as a defense against theater ballistic missiles.

C4I systems may also find new uses. For example, the Joint Surveillance Target Attack Radar System—originally acquired to serve Air Force and Army needs for ground surveillance—is now viewed as important to support Navy land strike missions.

Finally, in the course of using C4I systems in both exercises and deployments, users will often find a need for one system to interoperate with another when that need was not explicitly anticipated at the design stage of either system. Despite the best intentions and early planning of C4I system designers, there are likely to be many cases where user needs surpass those envisioned by the authors of the original requirements.

In all of these cases there is a need for architectural efforts across systems aimed at accommodating future unanticipated interoperability requirements.

BOX 2.4 Layers in the Open Systems Interconnection Reference Model

In the area of communications, the Open Systems Interconnection reference model[1] provides a useful framework for thinking about different levels or layers at which interoperability must be considered for C4I systems.

- **Physical**—connectors and signaling. Physical interoperability depends on things like electronic/photonic interconnectivity (the commonality of voltages and waveforms that allow communications over physical interconnections) and the mechanical compatibility of connectors. Interoperability problems that arise when two C4I system components have not been designed to share a common communications grid are also in this layer.
- **Data link, network, and transport**—the transmission of data (e.g., a message, binary representation of a picture, and so on) between C4I applications. For example, C4I applications that use different field radios with different framing and addressing conventions would not interoperate at the data link layer. To take another example, consider an application that produces and transmits data via a shared file and another application that expects to receive that data via a TCP stream. These two applications would not be interoperable, even if the applications agree completely on the semantics of the data and the battlespace models used to interpret the data.
- **Session, presentation, and application**—the interpretation of successfully communicated data. For example, two C4I applications may communicate position as three spatial coordinates but not agree on whether spherical or Cartesian coordinate systems are to be used, or which Earth model is to be used to interpret spatial coordinates, and thus they will not interoperate. More subtly, they will not interoperate if they do not agree on a common model of the battle space or have a means of translating between two different models (e.g., the scope and detail of information presented to a division command in a common operating picture vis-à-vis that presented to a tank commander, or the scope and detail of a common air picture for a naval air engagement vis-à-vis that required for a land engagement).

---

[1] International Organization for Standardization/International Electrotechnical Commission. 1994. *Open Systems Interconnection—Basic Reference Model: The Basic Model*, ISO/IEC 7498-1, International Organization for Standardization, Geneva.

Deployments Are Typically Conducted on an Ad Hoc Basis

Under today's arrangements, assembly of forces for contingencies is ad hoc, based on a generic set of requirements rather than preplanning that designates specific forces for a particular contingency. The number of possible force combinations that might be employed and for which interoperability must be tested, exercised, and established to meet the contingency is thus larger than would be the case if they were at least in part predesignated. This complexity also increases the difficulty of assessing the interoperability component of the readiness of forces that might be called on for a given deployment.

Coalition Warfare

Operations that involve coalition partners impose special challenges to achieving interoperability. With countries in a formal military relationship with the United States, such as North Atlantic Treaty Organization (NATO) members, there is an established framework in which to work on interoperability challenges. Absent a formal treaty relationship, as was the case with non-NATO coalition partners in the Gulf War, or when it is less predictable who the coalition partners will be, it is far more difficult to deal with interoperability challenges in advance of an operation conducted with a coalition partner. In both cases, but particularly in the less formal coalitions, the U.S. partners will not necessarily have adopted the same set of standards—even commercial ones—as those used by the United States. A final challenge is that coalition partners in both classes typically spend less than the United States on modernizing their C4I systems and thus may well be using equipment that is substantially less capable than and incompatible with present and planned U.S. C4I systems. The introduction to the recommendations in this chapter and section 4.3.4 in Chapter 4 discuss these and other challenges of coalition warfare for C4I systems.

Flexibility to Accommodate Variations in Command Structures

Command relationships determine the overall nature of information flows, and joint task force and theater commanders have the discretion, and the responsibility, to determine these command relationships as they think best. Commanders who choose unconventional command relationships may well require C4I linkages that have not been attempted before. New C4I linkages are likely to be required both on a large scale whenever a joint task force is deployed, as well as on a smaller scale even in an established theater (e.g., Korea) that receives new units or forces.

Flexibility to Accommodate Changing Tactical Situations

C4I systems need to be able to be reconfigured, augmented, and redeployed as the tactical situation changes and the various C4I components fall under attack, are required to support multiple operations at the same time, or are otherwise placed under stress.

Short Configuration Times

Particularly stressful are operations that are short-notice, "come as you are" events (e.g., Grenada). Here the requirement is that C4I interoperability be achieved within hours or days, rather than the weeks or months typically available in the commercial sector. Especially in operations when little time is available for field integration, interoperability limitations among C4I systems impose limitations on how systems can be connected and thus may impose undesirable constraints on what command linkages can be implemented.

Long Cycle Times for Complete Upgrades Leading to Heterogeneously Equipped Units

DOD procurement budgets usually make it infeasible to deploy a given generation or version of technology widely before that technology increases significantly in capability. Such budgetary limitations have two significant consequences:

- Because the time needed to equip an entire service is long compared to the time scale on which information technology changes (e.g., processor power improves by a factor of 10 in 5 years, but it takes on the order of 15 years to fully equip the Army with modernized tanks), the technology underlying a new C4I system will be much more capable at the end of the procurement cycle than in the beginning. Backward compatibility of new generations with preceding generations thus becomes an important issue to resolve.[11]
- While a new C4I system is being installed, and perhaps throughout the system fielding cycle, different units may have different capabilities.

---

[11]As discussed in section 2.2.1, backward compatibility is supported in the commercial sector for only a limited number of generations. Such practices are reasonable in the commercial world given the rate at which users are upgrading hardware. Whether the same time scales should govern military acquisition of C4I systems is less important than the underlying point—at some point determined by the underlying information technology (and not by the C4I system of which it is a part), compatibility efforts may have to be abandoned and incompatible upgrades may prove to be essential.

The division selected to be first equipped may have Version 2 of a system, while all others have Version 1. Or, the initial division will have Version 1 (as will happen in the Army's digitization effort), while the others have nothing. Then too, a given contingency may require digitized and non-digitized divisions to work together. These interoperability requirements—compatibility in doctrine, tactics, training, and ability to exchange information with non-modernized units—place many constraints on modernization efforts.

In short, the C4I systems in use in the individual services will not, and cannot, stay in lockstep. Thus, even a "perfectly interoperable" information system would not solve the problem permanently. The rate of change of both technology and warfare would ensure that such perfection would be at best transitory.

Building Horizontal Systems in a World of Vertical Organizations and Programs

Although the goal for many C4I systems may be that they be interoperable in a joint environment, this horizontal objective must be realized in a world that is fundamentally vertical. The vertical focus of system acquisition comes from two major sources. First, C4I system components are typically acquired by the services and funded out of service budgets. Second, the acquisition system itself is geared toward the development and procurement of discrete components rather than system-wide capabilities. Program managers are generally held accountable for the performance, cost, and schedule of their piece of a system, not for the performance of the whole system of systems.

2.3 TECHNICAL APPROACHES TO INTEROPERABILITY

As noted above, all large organizations have trouble achieving interoperability. As a practical matter, large organizations are generally not able to start with a blank slate with their information systems. In other cases, enterprises that are reengineering across a range of business practices, or that are restructuring operations, will introduce major new systems.[12] In both instances, it makes sense to strive for an information systems environment—perhaps never fully realized in practice—that is based on a clean architecture and requirements specification, common data structures, common interface requirements, and well-specified high-level

---

[12]For example, companies such as SAP—and a host of consultants and implementers—have built a multibillion-dollar business around replacing existing systems in support of reengineering efforts and mergers and other new combinations of business units.

information flows. Systems constructed in accordance with such an architecture are much more likely to be adequately interoperable than those that are not.

2.3.1 Architecture

Architectures are a hierarchical description of the design of a system and in many cases how it will be developed, evolved, and operated—"the structure not only of the system, but of its functions, the environment within which it will live, and the process by which it will be built and operated."[13] Architectures provide the underlying blueprint for the more detailed design and implementation decisions about components of a system. When well-defined architectures exist, engineers can design individual components and builders can implement them with a high degree of confidence that the end result will work as expected and meet user needs. Successful architectures are driven by more than technical consideration—they have as their fundamental goal the support of the requirements of users throughout an organization and are often represented in multiple dimensions, e.g., functional views, physical views, and operational views. When done well, architectures have enormous influence on the success of the overall endeavor. Some examples of commercial information systems architectures that have had such impact are the Ethernet local area network, the IBM S/390, and Digital's VMS.

Within an organization, development of architectures goes hand in hand with business process reengineering. In the military context, such business process reengineering would translate to an examination of how doctrine and procedures might evolve to exploit new capabilities offered by C4I systems (see also section 4.1.4). The mere automation of existing processes results not only in less-than-optimal gains, but also in "islands" of functionality (determined by the preexisting business processes) that exchange information only with great difficulty. Because reengineering requires an understanding of information flows that cut across old organizational boundaries, it lays the intellectual groundwork for an architecture that will support those flows.

2.3.2 Interfaces, Layers, and Middleware

Interfaces

Systems that perform a variety of functions are normally composed of multiple subsystems or components. Interfaces arise whenever one com-

---

[13]Eberhardt Rechtin. 1996. *The Art of Systems Architecting*, Prentice-Hall, Upper Saddle, New Jersey.

ponent or subsystem needs to interact with another. Principles for the partitioning of systems and selecting interfaces include:

• *Managing modular development.* Partitioned system design with well-designed interfaces permits development programs to be divided into more manageable pieces, which in turn can result in faster development because the work of different players can proceed in parallel. When it is desired to spread system development across different operating units, this approach is essential.

• *Permitting modular change—in versions and implementation technology.* By encapsulating the internal details of a system component (which may change over time), interfaces allow changes in internal implementation of portions of a system to be transparent to other portions. Interfaces thus permit various parts of the overall system to evolve over time without requiring changes to be made simultaneously throughout a system—allowing components to be upgraded as technology evolves and user needs mature.

• *Reducing the number of interaction points between systems.* Reducing the complexity of intersystem dependencies facilitates more rapid reconfiguration of systems to meet operational requirements.

The modular decomposition of systems often is both horizontal and vertical. Vertical decomposition refers to interfaces between discrete systems within the same layer—for example, a standard message format used by two different applications to exchange information. Horizontal decomposition of functions in an architecture—for example, the separation of bit transport technologies, transport protocol, and applications—is known as layering.

Layering

Layering can do a great deal to facilitate making C4I systems interoperable in the presence of rapidly changing technologies and multiple technology choices. Layering makes it possible to design a system of systems that has technology independence, scalability, decentralized operation, appropriate architecture and supporting standards, security, and flexibility, and can also accommodate heterogeneity, accounting, and cost recovery.[14] When layers in a system are identified, it is important to realize that layers that correspond to widely adopted standards are likely to be the most successful. The wide popularity of the Internet and the prolif-

---

[14]See Computer Science and Telecommunications Board, National Research Council. 1994. *Realizing the Information Future: The Internet and Beyond*, National Academy Press, Washington, D.C.

eration of Internet applications are evidence of the power of this principle. Specific examples include:

- *The use of TCP/IP to decouple communications link technologies from applications that use communications.* A diverse, rapidly changing set of link-level communications technologies (ranging from analog telephone circuits to wavelength division multiplexed optics) is thus separated from an even more diverse set of applications. System designs based on this layering principle produce applications that can interoperate with each other through networks built from a variety of technologies.
- *The use of hypertext transport protocol (HTTP) and hypertext markup language (HTML) to separate presentation from storage and retrieval functions.* Presentation on a variety of client platforms is separated from a rapidly evolving, wide variety of server-side applications and functions.

Middleware

Middleware is one instance of the layering principle. It provides a separation between applications and the operating system platforms that the applications run on. By decreasing the dependence of applications on a particular operating system, middleware increases the ease of moving applications to new computer systems and decreases dependence on operating systems that might fall out of favor in the commercial marketplace.

Middleware has an additional dimension—as a toolkit, it provides a set of relatively high-level common functions that are used in common among applications, and permits applications to be built out of building blocks. Examples of functions that can be provided by middleware are file system support, privacy protection, authentication and other security functions, tools for coordinating multisite applications, remote computer access services, storage repositories, name servers, network directory services, and directory services of other types.[15] Thus middleware offers two additional advantages. First, when common software is used to provide particular functions, interoperability is more easily achieved. Second, by increasing software reuse, middleware can reduce development costs.

2.3.3 Standards

A key element of architecture is the establishment of technical standards. Such standards define common elements, such as user interfaces,

---

[15]Adapted from Computer Science and Telecommunications Board, National Research Council. 1994. *Realizing the Information Future: The Internet and Beyond,* Chapter 2, National Academy Press, Washington, D.C.

system interfaces, representations of data, protocols for the exchange of data, or interfaces accessing data or system functions. Examples of standards include UNIX, Windows, TCP/IP, structured query language, and the Defense Information Infrastructure-Common Operating Environment. Technical standards offer a number of advantages for a system architect:

• They make it easier to exploit changing technologies (Box 2.5). For example, standardized interfaces facilitate interoperability because a component that conforms to a given standard can "plug" into a standard interface without concern for how the component on the other side of the interface works internally.

• They provide an understanding of data or a platform that is common to all component developers.

• They facilitate interoperability because they are accepted by multiple vendors and thus increase the likelihood that a collection of systems from diverse sources will be able to interoperate.

As advantageous as standards are, an approach to interoperability based on standards and/or standards-compliant common products must deal with certain realities and issues:

• In some areas, capabilities or services of interest are not covered by standards, although de facto standards, instantiated in broadly used products, can be an attractive option.

• In other cases, there are standards but no existing, mature products (e.g., standards "holes" in functional areas or relative to features such as security).

• Even when there are accepted standards and products compliant with these, interoperability is facilitated but not assured; there are options within standards, different releases and versions of products, and so on. *The devil of assuring interoperability is in the detail of implementation.* For example, the definition of an interface standard might not specify allowed and disallowed sequences in which a connecting component may call on different system functions. Thus, a component that issues a particular sequence of calls may cause a malfunction in the other component if that sequence was not properly anticipated. Vendors may also add additional capabilities or features to distinguish their products from those of their competitors; systems that rely on these features may not interoperate with systems that more closely follow the standard.

• There is a natural tension between adopting standards and taking advantage of the continuing stream of improved capability offered by technology, now dominantly driven by the commercial marketplace.

• Finally, it is important to realize that technical standards are, by themselves, necessarily incomplete from the standpoint of a system or

> **BOX 2.5 Interface Standards and Rapid Exploitation of Technology**
>
> Information technology is characterized by rapid change. How can such change be exploited by the system designer?
>
> One approach to technology exploitation is to rely on standardized interfaces so as to avoid the need for tight "vertical" integration of system components. Consider, for example, a system consisting of a set of sensors providing input to a set of databases, on top of which is built a system providing data integration, analysis, and decision support. Progress will be made in both the front-end sensor technologies and in the (mostly commercial-off-the-shelf) technologies supporting the back-end analysis and decision support, but this progress may be made at very different rates. Particularly in time-critical applications, it may be the case that frequent upgrades to specific functional capabilities in decision support could pay huge dividends. When care is taken to establish a well-defined interface between the sensors and the databases, and another interface between the databases and the decision support tools, different parts of a system can develop at different rates independently of each other.
>
> It is true that designs that rely on standardized interfaces cannot take advantage of special characteristics of the components themselves, with the result that an interface-based design may have poorer performance in some dimensions (e.g., speed, bandwidth) than a tightly integrated one. Tight integration historically has characterized military systems. For example, in the Joint Tactical Information Distribution System the message formats, the waveforms, and the hardware are highly intertwined. But in addition to not allowing exploitation of all of the good properties of layering (e.g., minimal interaction between layers and thus greater ease in "debugging"), such an approach ignores the fact that a tightly integrated design must—by assumption—proceed at the slowest development pace that characterizes any of its components. In a world in which the underlying technologies evolve so rapidly, the performance benefits of tight integration come at the cost of not being able to use new technology as it becomes available—on balance, a losing proposition.
>
> Well-defined interfaces also enable the creation of reasonably accurate system models for use in optimizing a system and understanding the performance enhancements that will result from specific localized upgrades.

component designer. The important thing is the operational scenarios that a system is expected to support. This range of scenarios defines the context in which a system is to perform specific desired functions and thus provides a meaningful reference for testing and evaluation.

As a general rule, some standards gain widespread acceptance in the commercial computer and communications industries and thus tend to have a long lifespan. The marketplace tends to weed out weak standards before they become widely accepted. And once a standard is widely used, industry is often motivated to maintain compliance with this accepted standard. Standards created by niche players in the market tend not to survive.

DOD, despite its size, is a small force in the overall marketplace, which suggests that if DOD attempts to create its own standards, the standards will not be viable in the long run except where they are relevant only to military applications and do not have to compete with analogous standards in the commercial sector. DOD is more likely to be successful if it exploits well-articulated and tested commercial standards wherever possible in C4I systems. An example is the use of TCP/IP. Although TCP/IP lacks certain features that would be helpful in the military environment, it is widely and successfully used by DOD. Even in those cases where today's commercial technology is not sufficient to satisfy DOD needs, a DOD-specific development is not necessarily justified. It can also be useful to project where commercial technology will be, in terms of its capabilities, in the time frame in which a DOD-specific product would realistically become available.

However, deficiencies in the security of many commercial technologies represent a special case and deserves special attention (see Chapter 3). Frequently, the best approach is to accept an 80% compromise solution (see section 4.3.2 in Chapter 4) that meets the bulk of DOD requirements.

When essential capabilities are missing from a standard, the best approach to dealing with the shortcoming is to either work to extend the standard or develop extensions to the standard. Because the use of commercial standards is such a powerful tool for ensuring ongoing interoperability, supportability, and upward evolution, designers may be well advised to use such standards even when they may lack certain features. If the lack of certain features is critical, then a custom development or an effort to have such features included in the commercial standard may be necessary. In other cases, work-arounds may be possible that enable the final product to meet the vast majority of its functional requirements. Program managers must have both an explicit strategy for developing such work-arounds as well as a credible analysis that indicates that the use of a commercial standard is tolerable. Section 4.3 discusses approaches to the use of commercial technology.

2.3.4 Data Interoperability

Experience suggests that left to their own devices, the designers of individual systems will often make locally optimal decisions about data definitions and formats. Data formats resulting from such local decisions may not be compatible when operational requirements dictate that a network of systems be called upon to interoperate. For example, different applications might use the same field code to mean different things, thus leading to interoperability problems among the applications. Thus architectural design must provide guidance to developers to minimize the applications-layer incompatibilities that inevitably arise when systems with different purposes must communicate with each other.

Examples of approaches to data interoperability include:

- *Single data definition for all systems.* Mandating a single data definition for all systems is conceptually appealing. However, it is dangerous for several reasons, especially when applied on a large scale to a complex, evolving system (or system of systems):
    — Across a large organization (in which subunits have different needs and view the same concepts differently), it is a very difficult task to agree on definitions. The definitions that emerge are likely to have mistakes. Other approaches also may result in errors, but because their scope is limited, the errors have limited consequences.
    — The task of agreeing on definitions consumes a great deal of effort and time that might be better used elsewhere. Waiting until agreement is reached may be very costly.
    — Even assuming that a single set of data definitions can be developed for a set of applications, it may well be difficult to design a new application around those definitions. And developers of later applications do not have the benefit of having helped to develop the initial definition.
    — When a single set of definitions is mandated for all applications, definitions are no longer locally optimal, and thus such mandates often encounter substantial resistance in implementation.
- *Object orientation.* A technically promising approach is to use object-oriented concepts to develop data definitions, encapsulating the internal details of the data. A change in the representation or definition of the data then has minimal impact on the applications that access that data. Contrast this with a linear record that defines some data object—the result is that all the applications that access the data are subject to change when any changes are made in the record.
- *Extensible data model.* Another approach to achieving data interoperability uses an extensible data model and standardized interface. The Simple Network Management Protocol is a good example. It has a generic description of types of objects that can be extended by applications

without requiring universal agreement on the extensions. This approach allows new devices to interoperate with the rest of the network management system.

Legacy systems, which have been built around frequently unique data definitions, pose a major challenge to interoperability. Industry has developed a number of approaches by which systems not designed up-front for interoperability can interoperate to exchange information:

- *The data "bus" approach.* Each system uses its own data definitions internally. However, exchanges of data with other systems are conducted through a "bus," that is, a common data standard into which data must be translated before being transmitted to another system. Any system wishing to use this data then downloads it from the "bus" and retranslates the data into locally meaningful terms before that data is used.
- *The data dictionary approach.* Each system has a published data dictionary and a simple query-response mechanism to access the data with published message formats. Given a later need to interoperate, another supplier could build to that embedded base interface and access the system's data. A system with this capability may cost marginally more than a closed system, and additional security issues need to be addressed, but it is often a reasonable poor-man's approach to interoperability when nothing better exists.
- *The data translator approach.* Two systems that need to interoperate have a translator that converts one set of data definitions into the other. This approach preserves the internal integrity of a legacy system, but the translators may be slow and, more important, may not preserve the original semantics of the underlying data.
- *The data server approach.* Data and processing are separated. When a system requires data, it connects to a data server that provides the data. Thus, enforcement of definitions can be limited to just a few servers rather than a myriad of applications. By moving the data into a system separate from the individual applications, this approach facilitates reuse of data in new, unanticipated ways.

2.3.5 Developing and Implementing Architectures

Commercial best practice suggests the following principles for developing and implementing architectures.

Use a Small Architectural Team

A critical element of developing a good architecture is the involvement of a good architect. The role is demanding, requiring an ability to balance needs and resources, technologies, and the interests of multiple stakeholders. As a general rule, good architecture results from a small number of people (perhaps a single chief architect, perhaps a small, closely knit team) being responsible for its content and structure. Architecture is both an art as well as a science, and a good architect must rely in part on aesthetics. Good architectures are less likely to emerge from large teams or from a broad consensus-based approach because the search for consensus almost always results in unwieldy compromises that have negative effects on system building. Consensus architectures are likely to lack a clear design philosophy, which often causes confusion among implementers.

Position the Architect Appropriately Within the Organization

Developing an architecture is an endeavor that touches on units throughout an organization, and care must be taken to position the architect. To maximize the chance of success, the architect must:

- Not be owned by any subset of the organization,
- Have the support of the top leadership,
- Have an appropriate charter and sufficient authority, and
- Have sufficient technical resources.

Limit the Scope

The perils of attempting to establish architectures over too wide a scope have been seen in a number of instances, and caution is in order in approaching C4I interoperability with a goal of a fully integrated C4I system of systems with seamless interoperability. One should not forget that the world has seen more than a few failures associated with global efforts to reshape the software landscape over a short period of time. Examples that come readily to mind include DOD's effort to establish ADA as the universal programming language (Box 2.6). The lesson to be learned is that in the hope of achieving a grand and universal solution, it is easy to grossly underestimate the complexity associated with a project of such large scale, the difficulty of managing it, and the level of talent required at all levels to achieve success. Indeed, scope must be limited for several reasons, including overall complexity, the need for scale to be commensurate with the pace of change in both missions and technologies, and the need to use small teams to develop good architectures.

Foreseeing all of the kinds of applications and their combinations is something that is both very difficult to do and often not successful. Before its breakup in 1984, the Bell System network was a good example of a

BOX 2.6 Commercial Failures Arising from Excessively Large Architectural Scope

Open Systems Interconnection

The International Organization for Standardization's Open Systems Interconnection (OSI) standards were an ambitious attempt to standardize networking protocols from the network level up to the application level. A great many committees met over a number of years during the 1980s and early 1990s and defined standards for packet formats, transport protocols, e-mail, directories, security, management, and many other topics. Many vendors, governments (including that of the United States through its Government OSI Profile program), and large corporations declared their intention to replace their use of various proprietary standards (such as DECnet and Novell), as well as the Internet protocols, with Open Systems Interconnection standards. But the whole enterprise came to nothing. All that remains in the market is a fragment of the X.500 directory standard called LDAP, and the X.509 public-key certificate standard. Everything else has been overwhelmed by Internet standards. Some of these, like the basic TCP/IP protocol and the Simple Mail Transport Protocol (SMTP) protocol for e-mail, had the advantage of being fielded before the Open Systems Interconnection effort started. But others, like the Simple Network Management Protocol (SNMP) and the Domain Naming System (DNS) directory service, came along well after the corresponding Open Systems Interconnection efforts, and usually in reaction to their excessive complexity.

ADA

Originally conceived as a single computer language that would be suitable (and mandated) for nearly all DOD programming efforts from financial accounting to real-time systems control, Ada's scope grew to quite unwieldy proportions and its use was often resisted even within DOD. Responding to the recommendations of a National Research Council report,[1] the DOD adopted in April 1997 a policy that eliminated the mandatory requirement for use of the Ada programming language in favor of an engineering approach to selection of the language to be used. Programming language selections would be made "in the context of the system and software engineering factors that influence overall life-cycle costs, risks, and potential for interoperability."[2] Thus, it is likely that programming languages that are more commercially viable and popular will be used to a much greater extent for DOD systems that are not associated with weapons systems or C4I systems.

---

[1] See Computer Science and Telecommunications Board. 1997. *Ada and Beyond: Software Policies for the Department of Defense*, National Academy Press, Washington, D.C.
[2] Assistant Secretary of Defense for C3I, April 29, 1997. *Assistant Secretary of Defense Memo: Use of the ADA Programming Language,* Department of Defense, Washington, D.C.

fully integrated system structure.[16] In this case, system integration was achieved over an extended period of time by strong top-down coordination, specification, testing, and strong management. In contrast to the public switched telephone network, however, the entire military C4I system of systems is far too complex and its missions change too rapidly for an approach of a single overall system design to be feasible, suggesting an approach to interoperability that focuses on more narrow mission areas. Overly tight integration also makes it more difficult to fall back to more independent operation when C4I systems are placed under stress.

Some would also argue that the Bell System was very slow to change with advances in technology because of its tremendous level of integration. Similar arguments can be made about the computer industry. Rapid changes in technology, cost, and features have often come about when the computer industry ceased to be vertically integrated. De facto interfaces were defined by the marketplace (e.g., instruction sets, operating systems, structured query language) that allowed intense competition and rapid, independent development on either side of an interface that no longer needed slow, centralized coordination of all of the parts of the network. Thus, it is necessary to strike a balance between striving for fully integrating systems, which brings with it a high degree of interoperability but likely will stifle quick innovation, and adopting a less constrained environment that permits faster exploitation of technological advances.

Finally, if the principle that architectural teams must be small is to be followed, the scope of the architecture must be limited. When a small team develops an architecture for a more narrowly defined operational scope, it is more probable that a well-designed architecture will result.

Engineer for Flexibility

Engineering for flexibility, so as to increase the likelihood of interoperability over time, includes several approaches discussed above:

- *Bias toward use of COTS.* Development of an architecture should rely as much as possible on the commercial market, and system designs should be based on compositions of COTS components. Then a substantial burden for interoperability, as well as continued development of the components, is passed to the commercial sector. Strategies for using COTS are discussed in section 4.3.3. This approach depends upon an acceptance of the "80% solution," discussed in more detail in section 4.3.2.

---

[16]Note that some historians of the telecommunications infrastructure argue that this integration was driven by government regulation. The market, circa 1985, was not interoperable and did not want to be—it was dominated by AT&T, which did not want to interconnect with the more than 6000 small networks.

• *Use of standards.* Use of technical standards is one way of planning for the future. Compliance with technical standards is an investment that facilitates, though by no means guarantees, future interoperability. Interfaces are another investment in the future; by providing well-defined ways of accessing systems and capabilities, they facilitate components composed in new ways in the future, or new uses of existing systems.

• *Investment in metadata.* Another investment in future interoperability is the use of sufficient metadata to enable data collected or generated by a system to be used in future applications in ways beyond the original intent. For example, providing geo-location data along with imagery makes it easier to use the imagery in a wider variety of applications.

2.4 TESTING

As argued above, an essential underpinning of C4I interoperability is architecture and requirements specification. Ensuring that the architecture and requirements are in fact successfully implemented, and that the required level of interoperability is achieved (which is not guaranteed by conformance to specifications), requires comprehensive testing and evaluation. Testing is critical to achieving interoperability and has an especially large payoff if conducted concurrently with development. Many interoperability problems are subtle, manifesting themselves only in certain sets of circumstances, and so are hard to uncover, and they demand a great deal of empirical work and testing to resolve.

Testing compares actual performance with requirements. It can take place in a laboratory, a field location, or at someone's desk with early system designs. Typically, systems are tested at different stages in their life cycle: during development, preproduction, and in the field (Box 2.7 describes DOD's efforts in these areas):

• *Developmental testing* assesses progress in meeting system-level requirements ranging from functionality to performance (including software stability). To ensure correct intent, a system's "paper" requirements may be tested against user-stated needs. Systems may be tested against requirements to ensure correct architecture and design. Subsystems may be tested against designs to ensure correct development.

• *Preproduction testing* is undertaken when a system has completed the development process but before it has been accepted for production.

—*Conformance testing* focuses on the stand-alone functionality and performance of a particular system. Through a paper or laboratory test, it validates the system in terms of stated requirements or specifications. The result of conformance testing typically is formal certification of compliance with the relevant standards.

BOX 2.7 DOD Testing and Evaluation

The DOD maintains an extensive test and evaluation structure that encompasses developmental and preproduction testing by the services' program offices and independent testing by designated service organizations reporting up through their service chiefs and to the Office of the Secretary of Defense's Director of Operational Test and Evaluation. The primary purpose of this testing is to ensure that a system meets specified functional and technical performance criteria and is operationally capable. The goal generally is to ensure that a system meets the requirements established for the system in its Test and Evaluation Master Plan, prior to its certification for full-scale production (as opposed to low-rate initial production) and its subsequent fielding and use by the operating forces. For C4I systems interoperability, the Defense Information Systems Agency through its Joint Integration and Engineering Office and subordinate Joint Interoperability Test Command performs operational test and evaluation for joint C4I systems throughout the entire system acquisition and deployment process.

Additional follow-on test and evaluation of C4I systems are also done for selected critical systems. This type of testing takes two forms. The first is a continuing test program of quantitative measures of the day-to-day operational performance of fielded systems. Diagnostic evaluations are performed to identify problem areas, and recommendations (concerning engineering or software changes, as well as procedures) are provided to address performance problems. Continuing follow-on testing and evaluation provide the operational and administrative commands a timely assessment of system operational performance and readiness. The second type of follow-on testing and evaluation for interoperability involves selected joint force exercises and tests in simulated operational environments. It provides both qualitative and selected quantitative assessments of the performance of C4I systems and is usually done at somewhat less than full scale, compared to actual operational environments.

Today, commercial suppliers are commonly regarded as having the primary responsibility for ensuring conformance to customers' requirements, transforming conformance testing from an adversarial test conducted by the purchaser into a more cooperative process (Box 2.8).

—*System-to-system* testing determines how well a system interoperates with other systems. It is typically performed in a laboratory, where two or more systems can be interconnected. Involving multiple systems and suppliers, it is usually more complex and expensive than conformance testing. Its scope can range

> **BOX 2.8 New Testing Relationships Between Vendors and Customers**
>
> Today, testing is generally performed comparatively early in a product's life cycle, as an integral part of the development process, and is led by the supplier with input from, or even the active participation of, the users. The supplier openly shares test results with its customers, thus minimizing the need for customer-performed conformance testing.
>
> Customers view suppliers favored in this way as strategic and often have risk-sharing financial relationships to maintain their interest, performance, and trust. Cooperative relationships often mean that suppliers understand customer needs better, time to market is shorter, and overall testing costs are lower. The disadvantage is that the customer may lose the additional level of assurance that a supplier product conforms to specifications.
>
> Despite the power of a more cooperative testing position, this type of supplier responsibility has typically not extended yet to end-to-end performance of systems interoperating with many other systems from many other suppliers. Achieving and maintaining end-to-end interoperability are often still activities for the customer/user to manage.
>
> An important corollary of having suppliers accept responsibility for the conformance of their systems to their customers' requirements is that this responsibility does not stop when a system is first fielded. Latent faults may not be discovered until new systems are later connected to this embedded system or the system is placed in some new environment. Suppliers practicing good quality management techniques accept the responsibility for later fixes to their systems. Their costs for performing this function either were allocated as an internal reserve of the original system purchase price or are recovered through customer-purchased "maintenance releases" of system improvements.

from "lower-layer" (e.g., communications) to "higher-layer" (e.g., applications/data) interoperability.

- *Field testing*[17] assesses the extent to which a system satisfies users' operational needs in a "real-world" setting, which differs from the controlled environment of developmental and preproduction testing: system configurations in the field (e.g., software releases, intermediate communications, etc.) are quite likely to be different, in detail, from the ideal configuration envisioned in the system design; those personnel operating the systems are typical field personnel rather than technically trained engineers; and nuances of system usage—often not apparent until a system is

---

[17]In DOD parlance, operational testing.

fielded—will arise, especially under non-ideal scenarios. Field testing is also essential because end-to-end interoperability involves critical nontechnical dimensions such as people, procedures, and training. Additional complications that require field testing to resolve may arise because corporate or organizational information systems are typically systems of independently developed systems (or components) in which unsynchronized component insertions can alter the interoperability properties of the overall system. Field testing involves functional testing and follow-on testing:

- *Functional testing,* the initial test in the field, cannot occur until people in the field have been trained in both the system and the business processes that the system will support. Functional testing involves configuring systems to meet the unique demands of particular customers, integrating products with the embedded base of systems (including earlier generations of the same product), and evaluating the resulting system of systems from the end-to-end functional perspective of the user.
- *Follow-on testing* assesses a system's performance after it has been fielded, reverifying interoperability periodically or as changes occur and providing a mechanism for tracking progress in addressing known problems. Some requirements cannot be adequately tested during the functional testing phase, and are best assessed during ongoing operations. Follow-on testing draws on information from multiple sources, including problem reports and lessons learned in joint operations and exercises, vendor information about features and bugs in new releases, and periodic monitoring of system performance and failures during field use.

In an ideal world, with an absolutely complete set of interface requirements and complete exercise of each system, conformance testing would catch all possible flaws. However, requirements are seldom complete enough to allow thorough testing, and complete testing takes too long. Often, requirements are strong in specifying behavior under ideal (sunny-day) conditions and weak about what should happen when it rains—for example, what the response of a system to a failure somewhere should be. System-to-system and field testing compensate by testing actual systems under a variety of conditions that go beyond those typically stated in requirements.

Testing should also be seen as an integral part of requirements definition and system development. Particularly in functional and follow-on testing, the value comes as much from having a process for learning about new requirements and feeding those requirements back from the operators to the developers as from identifying and correcting mistakes. As spiral development (see the discussion of evolutionary acquisition in section 4.3.2) becomes the normal mode for acquiring C4I systems, such mechanisms for rapid feedback become especially important.

Thus testing must be essentially continuous, and "stability" is a state that is never reached in any meaningful sense. Only when information is fed back to system developers and maintainers can processes and systems be modified to help ensure continuing high performance as the operating environment changes. Without ongoing feedback, initial implementations of processes and systems may interoperate satisfactorily at first, but not later.

2.5 DOD INTEROPERABILITY STRATEGY

2.5.1 Overview

Historical approaches to interoperability by the DOD have ranged from dealing with interoperability issues program by program to making limited-scope efforts on a joint, community-wide basis (e.g., the Joint Interoperability of Tactical Command and Control Systems activity to address joint message standards) or a functional community basis (e.g., air defense). In addition, some programs to develop defense-wide infrastructure, dating back to at least the 1960s, have been followed more recently by a few sizable, centrally managed application development programs (e.g., the Global Command and Control System as a replacement for the Worldwide Military Command and Control System).

In recognition of the leverage afforded by C4I and the importance of interoperability in realizing this leverage, over the last 3 or 4 years a more centralized, inherently joint/defense-wide strategy for promoting interoperability has emerged, comprising two major elements: a triad of interrelated architectures and a common defense-wide infrastructure, the Defense Information Infrastructure (DII) with a common applications platform, the Common Operating Environment (DII-COE, discussed in more detail below) as a key ingredient.[18] Responsibility for interoperability is distributed across DOD, and each of the major players has at least one entity charged with responsibility for interoperability issues. For example, the Defense Information Systems Agency has the Joint Interoperability Test Command, the U.S. Atlantic Command has the Joint Battle Center, the Joint Staff has the Military Communications and Electronics Board, and the Assistant Secretary of Defense for C3I has the Information, Integration, and Interoperability Directorate. Today, DOD is just at the begin-

---

[18]The committee learned about this strategy through various briefings, discussions, and site visits.

ning of refining and even establishing the processes and organizations to respond to future needs for C4I interoperability.

2.5.2 Elements of the DOD Strategy

Architectural Triad

The Defense Department has defined three interrelated architectures for C4I systems: the Joint Operational Architecture, the Joint Systems Architecture, and the Joint Technical Architecture (Box 2.9). The Joint Operational Architecture is intended to identify mission objectives, information exchange requirements, and logical connectivities among and within command and control units or organizations. The Joint System

---

BOX 2.9 The DOD Three-Part Architectural Framework

OPERATIONAL ARCHITECTURE—"a description (often graphical) of the operational elements, assigned tasks, and information flows required to support the warfighter. It defines the type of information [exchanged], the frequency of the exchange, and what tasks are supported by these information exchanges." The operational architecture is thus the doctrine-driven representation of C4ISR nodes, roles, processes, interrelationships, and data/information exchanges. This representation relates to specific scenarios and joint/combined/coalition mission functions and forms the basis for realistic process and information flow representation and prioritization.

SYSTEMS ARCHITECTURE—"a description, including graphics, of the systems and interconnections providing for or supporting a warfighting function. The systems architecture [view] defines the physical connection, location, and identification of the key nodes, circuits, networks, warfighting platforms, etc. associated with information exchange and specifies system performance parameters. The systems architecture [view] is constructed to satisfy operational architecture component requirements per the standards defined in the technical architecture."

TECHNICAL ARCHITECTURE—"a minimal set of rules governing the arrangement, interaction, and interdependence of the parts or elements whose purpose is to ensure that a conformant system satisfies a specific set of requirements. It identifies system services, interfaces, standards, and their relationships. It provides the framework upon which engineering specifications can be derived, guiding the implementation of systems."

SOURCE: C4ISR Integration Task Force. 1997. *C4ISR Integration Task Force Executive Report*, Department of Defense, Washington, D.C., November 30, p. 27.

Architecture is intended to map these information exchange requirements to specific hardware and software systems and to specify capacity and performance constraints. The Joint Technical Architecture identifies and mandates standards and identifies standards-compliant products, when available, for the building of systems and subsystems so as to promote interoperability between them.

The architectures are not all at the same level of development; the Joint Technical Architecture (JTA) is by far the most mature of the three. Wherever possible, the JTA references commercial standards, products, and technologies. The JTA is intended to provide a set of correct and mutually consistent technical standards, application interfaces (APIs), and protocols, along with decision rules for using them. The scope of the JTA is broad, encompassing systems for C4I, sustainment, weapons and platforms, and modeling and simulation. By conforming to the standards, products, and implementing guidance codified in the JTA, such systems are intended to be "born joint," in accordance with Chairman of the Joint Chiefs of Staff Instruction 6212.01A.[19]

The Joint Technical Architecture also provides an important foundation for coping with unforeseen requirements. The investment in the basic level of interoperability that is offered by building systems in compliance with the Joint Technical Architecture establishes a defense-wide fundamental level of interoperability, which permits a much more rapid accommodation to new scenarios and operational requirements than would be possible without it.

As far as the committee has been able to determine, the Joint Operational Architecture was originally intended to be a construct covering all military operations. For example, the 1998 *Annual Report* of the Secretary of Defense states that "the DOD is developing an agency-wide Joint Operational Architecture that describes the tasks and activities, operational elements, and information flows required to accomplish or support the missions of the DOD."[20] The Information Superiority Campaign Plan of the Joint Chiefs of Staff calls for "the development of a high-level, C4 Joint Operational Architecture that integrates the joint warfare functions, from national level through operational level, into implementations of the JV2010 [Joint Vision 2010] operational concepts."[21]

---

[19]Chairman of the Joint Chiefs of Staff. 1995. *Instruction 6212.01A: Compatibility, Interoperability, and Integration of Command, Control, Communications, Computers, and Intelligence Systems*, Joint Chiefs of Staff, Washington, D.C., June.

[20]William S. Cohen. 1998. *Annual Report to the President and to Congress*, Department of Defense, Washington, D.C. Appendix K is available online at <http://www.dtic.mil/execsec/adr98/apdx_k.html>.

[21]*Information Superiority Campaign Plan*, J-6, Joint Chiefs of Staff, Washington, D.C., available online at <http://www.dtic.mil/jcs/j6/campaign/task3_1.html>.

DOD architectural development to date, which has focused on the Joint Technical Architecture—conformance to a "building code" and standards—is, as DOD recognizes, in and of itself insufficient to ensure technical C4I interoperability, since it fails to address some of the most important architectural elements required for interoperability. Many critical architectural elements, not yet developed, would be contained within the yet-to-be developed Joint Systems and Joint Operational architectures. Used in combination, these would define interoperability requirements to support operational mission information flows. For example, the Operational Architecture and Systems Architecture for a particular operational activity would define which service-developed systems would have to exchange what information over what media in what format.

The committee recognizes that development of the Joint Technical Architecture was a pragmatic first step to take, given that establishment of technical standards is much easier than establishment of an operational or systems architecture. The definition of information flows and data semantics required for operational or systems architectures is inherently complex, and additionally, provokes debate about how operations are to be conducted.

Common Information Infrastructure

A core element of the DOD strategy for C4I interoperability is the building of a common, defense-wide information infrastructure that includes but goes beyond traditional long-haul communications and associated services such as messaging. Box 1.3 in Chapter 1 describes a number of elements of the Defense Information Infrastructure (DII). The DII includes a set of common software, the DII-Common Operating Environment (COE), including increasingly capable middleware, on top of which service/mission-specific applications can be built. Software reuse/commonality is a key ingredient that is intended to reduce development time and cost as well as enhance interoperability. The DII, with the COE as a key element, is envisioned as a DOD-wide "public utility" that can be extended into theaters of operations for support of wartime as well as peacetime use.[22] Service- and mission-unique applications are to operate on a "plug and play" basis (e.g., software application program interfaces), with the common infrastructure providing basic capabilities and services. Use of the DII-COE and achieving compliance at certain levels is specified in the Joint Technical Architecture.

---

[22]Note, however, that this homogeneous common infrastructure constitutes a potential information security vulnerability. See section 3.2.5.

Data Interoperability

Applications-level interoperability depends in part on the capability for exchange of data among applications in a way that both preserves meaning and is mutually interpretable. DOD understands the importance of data integration and has over the years launched two major efforts in this area:

1. DOD Directive 8320.1, "The Enterprise Data Model Initiative," sets forth a DOD process through which standard data definitions in functional areas (e.g., C4I, logistics, health care) are developed and then subjected to a cross-functional review process prior to being adopted as DOD standards.[23] The goal of this process is to develop a complete set of standard data elements for DOD applications. The ultimate intent of the initiative is to bring all these data models together into one DOD-wide standard. One tangible result of this initiative is the DOD Command and Control Core Data model, which is now contained in the Defense Data Dictionary System managed by the Defense Information Systems Agency. Today, compliance with DOD Directive 8320.1 is mandated by the Joint Technical Architecture.

2. In contrast to the "top-down approach" of DOD Directive 8320.1, the Shared Data Environment (SHADE) program relies on a "bottom-up" approach. SHADE is intended to enable different C4I systems to share data segments (portions of databases, including those associated with legacy systems) and to use standardized access methods.[24] SHADE does not standardize data elements overtly. Instead, SHADE provides middleware for translating data elements from one system for another's use. If two systems have data elements with the same meaning (an assumption that must be tested in each case) and SHADE has a corresponding data element, then the middleware can transparently translate the data from one system though the common SHADE element and then back to the other system. SHADE presumes that, for reasons of cost and convenience, existing database segments will be reused and shared, and DOD data will increasingly reside in standardized, shared database segments. SHADE has demonstrated some success in enabling legacy systems to interoperate.

---

[23]Department of Defense Directive 8230.1-M, "DOD Data Administration," September 26, 1991.

[24]The Defense Information System Agency's *Defense Information Infrastructure (DII) Shared Data Environment (SHADE) Capstone Document,* July 11, 1996, is the basis for this discussion. This document and additional information regarding SHADE can be found online at <http://diides.ncr.disa.mil/shade>.

2.6 MEASURING INTEROPERABILITY

Evaluating current status or measuring progress in an area as complex as interoperability is difficult owing both to its multidimensionality (i.e., no single metric can possibly suffice to indicate the state of interoperability) and the difficulty in developing and applying precise metrics. Today, overall C4I status is generally introduced into readiness assessments as the judgment of a commander estimating his/her ability to accomplish his/her mission(s). An Army commander, for instance, generates a mission accomplishment estimate (MAE) with the status of C4I qualitatively considered along with the "measurements" of readiness for equipment, personnel, and training. The interoperability component of C4I readiness is particularly difficult to assess. For one thing, unlike other kinds of resources typically included in readiness reporting (e.g., personnel, equipment on board, equipment serviceability, training reported at the unit level), interoperability inherently cuts across units.

Although some qualitative assessments of the status of a unit's C4I systems, including interoperability, may enter into readiness assessments using today's process, the increasing importance of and reliance on C4I support of military operations suggest that the status and health of C4I—along with interoperability and other key aspects—be introduced as a more explicit and objective (rather than implied and subjective) factor. In the absence of precise metrics and recognizing multidimensionality, it is reasonable to use scorecard techniques based on human judgment to capture how well a unit (or DOD as a whole) is doing with respect to *technical* implementation compliance, *system-to-system* interactions, and *operational* mission effectiveness.

2.6.1 A Technical Compliance Scorecard

The technical view of an architecture focuses on the criteria governing the implementation of specific system capabilities or attributes. From an assessment perspective, the concern is whether a given system's implementations comply with the applicable standards and guidelines. A technical compliance scorecard could be viewed as a list of systems with pass/marginal/fail ("green"/"yellow"/"red") ratings of their compliance with the relevant standards and guidance (Figure 2.1).[25]

---

[25]Such an approach has been used by DOD as part of a structured process for describing and evaluating levels of interoperability vis-à-vis operationally driven requirements (the so-called levels of information system interoperability).

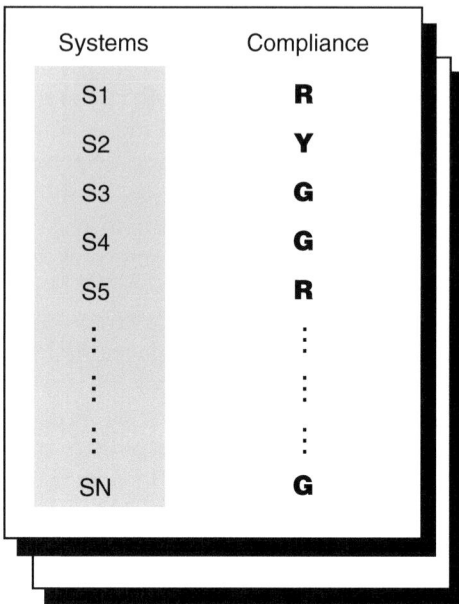

FIGURE 2.1 A technical compliance interoperability scorecard. The entries rate as pass/marginal/fail (green, yellow, or red) the compliance of systems S1, S2, ... SN with the relevant standards and guidance.

2.6.2 A Systems Interoperability Scorecard

The systems view of an architecture focuses on the information and communications systems that are brought to bear to support the information flows required to accomplish operational missions and attempts to measure the degree to which the various system pairs can effectively interoperate in context to meet these information flow requirements. These information flow requirements indicate the content and nature of the information and services needed, their directional flow, and the constraints and demands imposed by the operational environment. Accepting some oversimplification, one can view the problem as decomposition of a system architecture into a set of interconnected system pairs, which must each be able interoperate at some level of interoperability.

In this view, a scorecard used to measure interoperability from a systems perspective would derive from a codified (or de facto) system architecture, and would focus on the ability of the systems in each pair to interact with one another. Whereas in the technical compliance scorecard individual systems are assessed in isolation from each other, in the sys-

	S1	S2	S3	S4	S5	...	SN
S1							
S2	G						
S3	Y	R					
S4	Y	G	NA				
S5	G	G	R	Y			
⋮	⋮	⋮	⋮	⋮	⋮	⋮	
SN	G	Y	R	G	G	...	

FIGURE 2.2 A systems interoperability scorecard. The entries rate as pass/marginal/fail (green, yellow, or red) the pairwise interoperability of the systems indicated in the row and column headings.

tems interoperability scorecard two systems can be scored as being interoperable with each other in terms of the kind and level of interoperability needed.[26] The scorecard (Figure 2.2) could be viewed as a matrix with the systems represented in both the rows and columns and entries indicating system-to-system interoperability as inadequate ("red"), marginal ("yellow"), or adequate ("green").

2.6.3 An Operational (Mission-Enabling) Interoperability Scorecard

The operational view of an architecture addresses particular mission or operational slices, such as targeting, close air support, force sustainment, or the like, of a broader operational setting. Within each slice, it captures the players involved and their interactions, their functions, deci-

---

[26]The scoring must also take into account the diversity of system versions that are likely to be fielded at any given time. While version 2.1 of system A may interoperate with system B, version 2.2 might not.

sions, and actions, and the flows of information postulated to support their particular roles in achieving overall mission effectiveness. Operational architecture perspectives can be depicted using node connectivity diagrams, where the node-to-node connections are described in terms of information flow requirements indicating the content and nature of the information and services needed, its directional flow, and the constraints and demands imposed by the operational environment.

A scorecard used to assess interoperability from an operational architecture perspective would focus on the ability to satisfy specific node-to-node information flow requirements (see Figure 2.3) and the collective set of flows needed to satisfy a defined mission or mission slice. The assessed degree to which each flow requirement can be met can be scored using green/yellow/red ratings. These metrics are often derived from lessons learned through crises or exercises (observed events and anecdotal feedback). They deal, of course, with questions of interoperability and not with the difficult, higher-level topic of measuring mission effectiveness.

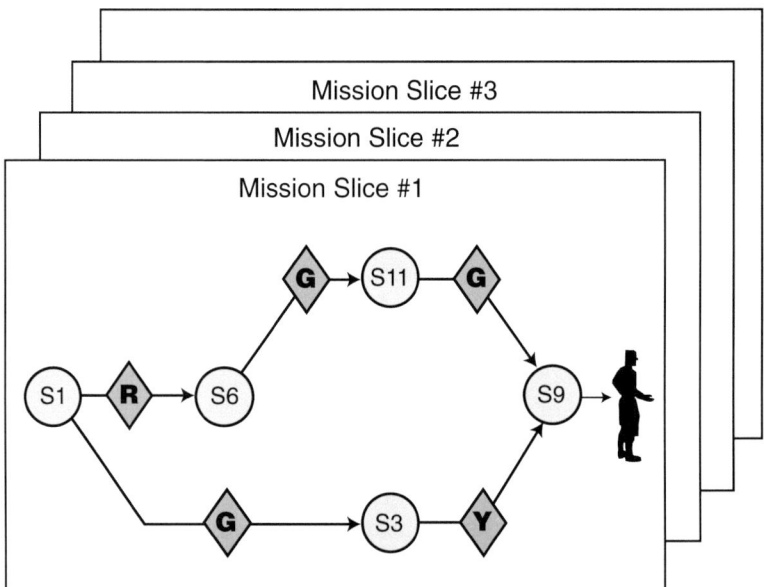

FIGURE 2.3 The operational (mission-enabling) interoperability scorecard. The diamonds rate as pass/marginal/fail (green, yellow, or red) the ability of the systems (indicated as circles) to provide the required information flows (indicated by arrows) for a particular mission slice.

2.7 FINDINGS

Although much has been done to achieve C4I interoperability, the goal of a C4I system of systems with assured interoperability for the U.S. military continues to be unachieved. DOD faces major challenges to assure effective exploitation of C4I. Progress has in many cases been slow, and past C4I studies show that many documented C4I interoperability problems remain unresolved.[27] Significant problems have occurred recently that have required significant time and resources to resolve (e.g., in the Gulf War; Bosnia).

The achievement of interoperability across a large-scale, complex C4I system of systems supporting military operations is a difficult undertaking. Despite increased attention and management awareness, along with a set of initiatives and strategies that the committee applauds, much more must be done before the infrastructure of C4I systems is as a whole largely interoperable and all new systems are sufficiently interoperable with the appropriate partners.

As discussed above, DOD has undertaken a number of efforts, at multiple levels within the services, the Joint Staff, the Office of the Secretary of Defense, and Defense-wide agencies, to deal with these challenges. Parts of DOD are well aware of a defense-wide problem in exploiting rapidly changing information technologies, in using COTS products effectively, and in assuring security. Today, a DOD strategy to promote interoperability exists, resting on the development and promulgation of technical standards such as the Joint Technical Architecture, the extensive use of middleware, and the evolution of a broader, enterprise-wide, common infrastructure.

Finding I-1: While the elements of DOD's current strategy for achieving interoperability are positive, they are not being fully executed. Both formulation and implementation have gaps and shortfalls.

Achieving interoperability in a changing world is hard. The DOD technical strategy (adopting an architectural approach, building to standards defined by the Joint Technical Architecture, and developing a common, defense-wide "public utility" infrastructure) builds on the best common practice in industry. It is a very important step that promises to significantly improve interoperability over time. At the same time, this

---

[27]See, for example, General Accounting Office. 1993. *Joint Military Operations: DOD's Renewed Emphasis on Interoperability Is Important But Not Adequate*, General Accounting Office, Washington, D.C.

strategy is not being fully executed. The committee highlights three areas in which execution has been deficient.

Lack of Progress in Development and Implementation of the Joint Systems and Joint Operational Architectures

The Joint Technical Architecture and its DII-COE and SHADE adjuncts are useful efforts. But they can only enable and facilitate interoperability, rather than assure it. The Joint Technical Architecture is only one part of the triad of operational, systems, and technical architecture. Furthermore, the Joint Technical Architecture and DII-COE address interoperability problems at the "lower layers" (e.g., data transport), leaving a large universe of "higher-layer" data and applications interoperability topics (e.g., data semantics) still on the table, though being worked on.

DOD is not fully executing its strategy in the formulation of the Joint Systems Architecture and the Joint Operational Architecture. The committee fully supports the fundamental idea underlying the Joint Operational Architecture, namely that obtaining the maximum value from C4I systems and networks depends on an understanding of how information is used by various parties in various circumstances (i.e., different operational scenarios) and how that information must flow between parties to support military operations. At the same time, the committee believes that the universe of all possible military operations is simply too large for a single Joint Operational Architecture to be developed successfully, and thus prospects for progress through a single DOD-wide operational architecture effort comparable to that of the Joint Technical Architecture are doubtful.

The committee did encounter some service activities, such as those of the Army Force XXI architect, that are seriously dealing with these issues, but did not see any visible effort to extend them into the joint arena. Significant progress on the Joint Systems Architecture was not apparent. A major reason the strategy is not being executed is that the DOD has defined an approach that is too broad in scope. For this reason, it is not surprising that Joint Operational Architecture efforts to date have mostly been void of operational content, with progress largely limited to definition of a framework and formalism.

One consequence of the lack of progress on the operational and system architectures is that the DOD strategy is thus far largely silent on security, except for security standards contained within the Joint Technical Architecture. One instance of this is the lack of architectural guidance for reconfiguring systems in response to cyber-attack. A common (and reasonable) response to increased levels of cyber-threat (Chapter 3) is for a system to drop non-mission-critical functions ("reconfigure the system" so as to reduce the number of avenues of information attack). However, in a visit to one exercise, the committee learned that mission-critical and non-mission-critical functions were not easily separated in the (implicit) Operational and Systems architectures of the operations center. Mission-critical functions were determined by a process in which all functionality is disabled and then individual functions restored (designated as mission-critical) when someone in the operations center demands them.

Security considerations cannot be managed in isolation. System architectures must deal with the issues of boundaries of trust and configuration controls relevant to that system. Operational architectures are intimately related to security because they specify information flows, and help identify those mission-critical functions for which information flows must be assured even under conditions of threat. Real-time determination of mission-critical functions while under attack is inevitably much more haphazard than a thoughtful consideration—included in the architectural design—of what functions are mission-critical at what level of threat.

Lack of Compliance with the Joint Technical Architecture

A major problem with the standards and common infrastructure approach has been a lack of compliance. For example, a recent DOD Inspector General report found that program plans for a large number of C4I projects did not call for conformance to the Joint Technical Architecture, even though those projects were subject to defense-wide guidance directing Joint Technical Architecture conformance.[28] The committee observes that if some written plans say nothing about compliance, it can only be assumed that others that do promise compliance will not in fact deliver. The report concluded that DOD does not have an integrated or coordinated approach to implementing the Joint Technical Architecture, and thus has little assurance that the Joint Technical Architecture will meet DOD interoperability goals.

A report by the General Accounting Office concluded that DOD organizations are not complying with the current interoperability testing and certification process for existing, newly developed, and modified C4I systems.[29] The General Accounting Office further found that in some cases, DOD program officials ignored the guidance, while in other cases, they were simply unaware of it.

---

[28]Office of the Inspector General. 1997. *Implementation of the DOD Joint Technical Architecture*, Report No. 98-023, Department of Defense, Washington, D.C., November 11.

[29]General Accounting Office. 1998. *Joint Military Operations—Weaknesses in DOD's Process for Certifying C4I Systems' Interoperability*, GAO/NSIAD-98-73, General Accounting Office, Washington, D.C., March 1.

These findings are consistent with the committee's sampling of C4I programs. Officials from the Office of the Assistant Secretary of Defense for C3I expressed to the committee their frustration at trying to persuade program managers to pay proper attention to C4I interoperability issues. Indeed, one person from that office argued that a major problem was that the definition of C4I technology and systems was porous enough that program managers could argue that they were not subject to directives applying to C4I systems.

Given that even with a formal process in place, Joint Technical Architecture compliance remains a major issue, the committee has concerns as to how system and operational architectures will be used. Indeed, it is unclear if an effective process is in place to use the system and operational architectures for development and fielding of systems.

The committee believes that the senior management of DOD recognizes many of the issues, such as the need for enforcement and providing sufficient resources for interoperability compliance efforts, but finds managing these issues to be extremely difficult. It is hard to enforce mandatory standards and guidelines across such a large organization. In addition, the term "mandatory" immediately raises the argument that there is legitimate mission-driven uniqueness, an argument that can only be addressed case by case. There is also concern about "unfunded mandates" except in cases where no additional cost is incurred, a rare situation.

The committee recognizes the limitations of an approach to interoperability that is based on enforcement. Effective enforcement of directives depends on enforcing bodies that have the authority to stop projects, resources to inspect the projects for which they are responsible, and willingness to exercise their authority.[30] Under such an arrangement, only a few programs can be influenced, and many others can escape the oversight and enforcement process. Nevertheless, particularly in the short run, there seems to be no viable alternative to enforcement as a management strategy. In the long run, interoperability will flourish only if DOD is able to promote a culture in which interoperability is valued, and in which individuals have strong incentives to build interoperable systems where required to support joint and defense-wide operations. Fundamentally, program managers must feel that their programs have failed unless they are interoperable. The Institute for Military Information Technology proposed in Recommendation P-3 (see section 4.8) would be an important venue for fostering a culture that values interoperability more highly, and

---

[30]In one instance where this authority was exercised, the Army Acquisition Executive acted to ensure that Force XXI systems would be built in compliance with the Army technical architecture.

would provide an environment supportive of the requisite cross-service collaboration.

Only Partial Success in Building and Using a Common Infrastructure

The DOD interoperability strategy rests in part on the use of a common infrastructure, the Defense Information Infrastructure (DII; see Box 1.3 in Chapter 1 for a description of some major common infrastructure programs), including a common software base, the Common Operating Environment (COE). In general, there has been insufficient migration toward the use of common infrastructure. Despite the commitment in policy toward use of a common infrastructure, there is still a proliferation of "stovepipe" systems.

Also, there is insufficient use of commercial products in the common infrastructure. Successful implementation of the DII-COE initiative requires attention to which functions should be DOD-developed and which should be taken from commercially available software. Also, the common infrastructure strategy requires careful attention to which operating system platforms the DII-COE will support. The common functions of DII-COE are implemented in a layer of software that is built upon an underlying operating system. The DII-COE must have a strategy for how it will manage multiple operating systems and evolve with the market for operating systems. The underlying operating systems should be COTS technology for a variety of reasons. These include the benefits of development investments based on a market much larger than DOD, testing by a larger community of users than DOD, maintenance costs that are borne by a larger base than DOD, functionality whose value is determined by the marketplace, and, most importantly, robust and creative COTS middleware and applications that are developed for high-volume platforms.

The committee observed that C4I systems today use a combination of UNIX and Windows operating systems. These represent the current choices for COTS operating systems. The DOD, like private industry, must monitor the market for these products, influence its direction, and respond to changes in its direction. For example, at present Windows enjoys the dominant share of the desktop, and a substantial and potentially growing share of servers. This market share, combined with other forces, attracts developers of middleware and applications that are of great potential value to DOD in C4I applications. A simple, but important, example that the committee observed in its field trips was the large number of office software suites for word processing and presentation graphics being used as part of the command and control process and providing important information flows.

The commercial sector, like DOD, must attempt to forecast and monitor industry efforts to replace the dominant product in the information technology marketplace, and adjust its COTS strategy accordingly. Examples of such attacks in the marketplace for application platforms include the Network Computer and Java, and Linux. The committee has no special insight into how dominant products in the market for operating systems may change. The committee does recognize that as changes occur, DOD must be committed to adapting the concepts of DII-COE to use the dominant operating systems to fully take advantage of COTS software.

In addition to having an operating system strategy for underpinning the DII-COE, DOD must continue to base middleware functions of the DII-COE increasingly on commercial products. For example, the committee sees Common Object Request Broker Architecture (CORBA) and Component Object Model (COM) as potential COTS replacements for certain elements within the current DII-COE. As with operating systems, DOD needs to monitor, forecast, and adapt to the market for middleware. The decision to adopt COTS components, an objective the committee endorses, should be based on an assessment of the trade-offs between the degree of support of specific DOD needs and the leverage afforded by using COTS. The committee sees this analysis as framed by determining when COTS does "80%" of what is required, and the impacts of adapting applications that use the DII-COE to a new middleware environment. The "80%" rule recognizes the benefits of replacing DOD development and support costs with purchase costs based on high-volume components, wider and more extensive testing, and benefits similar to those of the COTS operating system market (see above).

Finding I-2: Even full execution of the DOD strategy for interoperability will not assure that joint mission needs for C4I will be met.

Present efforts in the DOD interoperability strategy suffer from a number of weaknesses:

- *More must be done to prioritize interoperability needs and make the problem more manageable.* DOD efforts to construct a single Joint Operational Architecture are tantamount to specifying the information needs and requirements for all operations that the DOD believes it will have to conduct in the future. It would also have to cover an evolving set of C4I components and systems. Understanding the possible information exchanges between systems and components is at least an $N(N-1)/2$ problem (i.e., the number of possible pairs among N components). Because a single Joint Operating Architecture would require understanding how every part of a C4I system could be used in combination with every other part of any C4I system or component that is fielded, the unavoidable conclusion is that the ability to understand the entire system of systems does not scale well as components are added, and is clearly impractical.

On the other hand, an approach that depends on achieving interoperability on a pairwise basis is too narrow in scope. Because C4I systems are likely to grow in number and be synergistic and cooperative in their applications, a pairwise approach is unlikely to keep up with interoperability demands. These considerations suggest that the proper scope of a domain in which to address architectural issues is one that is more limited than "all military operations" but larger in scope than a pairwise system-to-system interaction.

The committee believes that a good organizing principle is that "proper" would be defined as a scope of analysis and concern that has operational significance, that is inherently joint, and that involves multiple systems. The same arguments apply to data standardization efforts and to the selection and use of tools that enhance data interoperability.

- *There is insufficient attention to building in interoperability throughout the life cycle of C4I systems.* Achievement of interoperability requires attention throughout development, testing, fielding, and deployment. The committee believes that current acquisition processes do not place sufficient emphasis on incorporation of interoperability during development. Once systems are fielded, a special emphasis is required on continued testing and verification of interoperability between systems. Even with such a regime of testing, interoperability problems will arise when units are actually fielded and systems are interconnected to meet the operational requirements of particular missions. Field support for commanders requires personnel who are knowledgeable and experienced in resolving interoperability problems, and who have a perspective that cuts across the full spectrum of C4I systems that interoperate.

- *There is no system in place to measure the interoperability state of C4I systems.* As discussed in Chapter 4, it is generally accepted that management must be able to measure what they wish to change. Today DOD management does not have in place measures of the current state of interoperability, either for assessing progress in developing and acquiring interoperable systems or for assessing the interoperability component of force readiness.

- *The strategy does not provide concrete guidance regarding technology evolution and the role of COTS technology.* Because the life cycle of C4I systems is long compared to the rate at which commercial information technology evolves, deployment is not an event occurring in a point in time but rather a process that takes place over years. Thus, architectures should provide guidance regarding strategies for deployment of the various components, both hardware and software, how they may be upgraded as the underlying information technologies increase in power and functionality, and most importantly, how upgraded systems will maintain interoperability with systems based on earlier generations of components.

- *Data interoperability efforts are inadequate.* Data interoperability standards referenced in the Joint Technical Architecture, such as the Defense Data Dictionary System, are mandatory defense-wide. But commercial experience suggests that because successful data models are based on an understanding of the interfaces in a system and how those interfaces are to be used, data models are more properly tied to operational and system architectures. Without them, an attempt at a data model will fail. Furthermore, an attempt at a DOD-wide data model seems doomed to failure as well—too many competing interests need to be coordinated, and it is likely that the effort will never converge.

The approach taken by SHADE is essentially the "data bus" approach described above. Such tools are potentially useful but require systematic application to operational mission areas for this potential to be realized. The committee's concerns are with regard to management and process rather than the technical approach. The SHADE effort depends on what amounts to voluntary adherence to a data interoperability regime. The committee understands the rationale for a voluntary regime, but remains concerned that persuading C4I program managers to use the SHADE approach will simply take too long to achieve a significant degree of data interoperability.

2.8 RECOMMENDATIONS

This chapter lays out a set of challenges that DOD faces in attempting to achieve a sufficient level of C4I system interoperability. The committee notes that some of these challenges stem at least in part from a broader issue, namely, the distributed, horizontal structure and organization of DOD itself, as established by Title X. In the recommendations that follow, the committee assumes no changes to this fundamental framework.

The committee's approach in making recommendations is to base them on principles and lessons learned from both the military and commercial sectors, and to focus more on outcomes than specific means. Thus the recommendations do not provide a high level of detail in identifying the specific ways to achieve these outcomes; these decisions will be dynamic in nature and are rightly made by the actors specified in the recommendations below.

Finally, the committee notes that its recommendations have applicability to the challenge of interoperability with at least a subset of coalition partners. Interoperability with partners that are unanticipated, and with whom no strong cooperative framework is in place, can largely be approached only through adherence to and influence on worldwide commercial standards. However, interoperability with partners who are members of an existing alliance framework or mechanism such as NATO or the U.S. military relationship with Korea, can be addressed by using approaches similar to those recommended below for dealing with joint interoperability: (a) the use of mission slices to focus architectural efforts, (b) the use of standards, particularly commercial ones, (c) a bias toward the use of COTS technology, (d) the scorecard approach to measuring progress in achieving interoperability, and (e) the establishment of a testing and field support activity. The committee recognizes, however, that the level of management complexity is clearly much greater when multiple nations are concerned.

Recommendation I-1: The Assistant Secretary of Defense for C3I and the Joint Chiefs of Staff should complement the DOD's current broad interoperability strategy with focused efforts in limited, operationally important domains, to include the development of Joint Operational and Joint Systems architectures for these domains.

The committee believes that it is more feasible to develop operational architectures developed for particular joint missions or tasks, organized around either significant operational capabilities or mission slices. When the spectrum of military operations is decomposed into joint "mission slices" and their supporting information "threads," the scope of the architecture problem and the data standardization problem both become more manageable. A mission slice is a component of an overall theater mission, such as close air support, suppression of enemy air defenses, or theater air and missile defense. Information "threads" supporting mission slices (e.g., the track files needed to support a theater air defense mission slice or the information flows associated with generation and execution of an air tasking order) can be identified and analyzed. These slices should not be confused with the sorts of vertical applications that were solved by what are today viewed as stovepipe solutions. Rather, they are intended as horizontal slices across the services and specific C4I systems.

A focus on architecture development for mission slices and information threads has two major advantages. The first is that is it allows DOD to set priorities. Progress in interoperability will take years, and the interoperability problem will never be solved "for good." It therefore makes sense to focus efforts on the areas of highest importance to the DOD. The second is that by prioritizing its efforts, processes and tools and techniques developed for the first efforts can be applied to later efforts, making those efforts easier to manage and more likely to be successful. This recommendation is not motivated simply by recognition of resource limitations or the need to "learn by doing." It also reflects the committee's view that an all-at-once development of an operational architecture covering the entire span of DOD's operational requirements is infeasible.

Criteria recommended for selecting and defining a mission slice or operating capability as a focus for interoperability efforts include the following:

- *The mission slice or operating capability should have considerable operational significance.*
- *The mission slice or operating capability should be inherently joint and involve a large enough number of systems to warrant the effort.* Selecting such a slice or capability ensures that the architecture effort will be horizontal in nature, and thus resolve interoperability issues rather than create new stovepipes. For example, both theater air and missile defense and the single integrated air picture are inherently joint, as they involve a varied mix of sensors and weapons whose information flow requirements pass through multiple service boundaries.
- *The mission slice or operating capability should have metrics or end-to-end performance indicators that can indicate improvement.* For example, performance indicators for the effectiveness of air and missile defense have existed for a long time (e.g., percentage of attackers that penetrate the defenses). In systems that provide capabilities such as the single integrated air picture (SIAP), the number of reported air tracks in the systems for every real object in the air or other such data would serve as comparable quantitative indicators of performance.
- *The mission slice should be one in which significant foundational work has been undertaken.* One mission slice for which this is true is theater air/missile defense, an area that is highly significant operationally, is associated with serious force integration issues, and—not least—has substantial operational and system architecture work already done by the Joint Theater Air Missile Defense Organization (operational architectures) and the Ballistic Missile Defense Organization (systems architectures).

Significant operational capabilities are another useful focus for interoperability efforts. The committee believes that programs providing a common operating picture represent a set of good choices, because they have broad applicability (interoperability needs are rich), are contained (bounded), and support an operational capability. Work on perhaps 10% to 20% of the data elements would yield a major interoperability payoff, and thus the effort, albeit expensive, would be rewarded. A key information thread of the common operational picture (COP) is the data elements in the "track file"; this would be an appropriate first thread on which to focus attention. COP programs have been conceptualized to provide capabilities at several levels. For example, within the Global Command and Control System (GCCS), the COP provides a common operating picture at the headquarters level. Other examples of operational capabilities on which to focus include the common tactical picture (CTP) and the SIAP—both capabilities critical to achieving effective joint operations and status reporting up the chain of command.

It is understood that the lead for developing C4I architectures is a shared responsibility of the Directorate for C4I Systems of the Joint Staff (operational architecture) and Assistant Secretary of Defense for C3I (systems and technical architectures), with support coming from other DOD elements depending on which mission slice is selected (e.g., from the Joint Theater Air Missile Defense Organization if theater missile defense were chosen as the mission slice). Given the urgent need to develop an operational and system architecture to guide ongoing development, the committee recommends that the Directorate for C4 Systems of the Joint Staff and the Assistant Secretary of Defense for C3I select an appropriate mission slice and initiate an activity to develop operational and systems architectures as a mechanism for identifying and prioritizing interoperability needs and problems—both today and prospectively.

Again, the consideration of available foundational work is important and suggests that a candidate would be a collaborative Joint Theater Air Missile Defense Organization/Ballistic Missile Defense Organization effort on the theater missile defense mission—an activity that would yield specific results for a crucial joint mission and also serve as a pilot of the "mission slice" approach.[31] Also, the committee believes that it would be useful to draw on service efforts to establish architectures for guidance in selecting mission slices and management approaches.

Note that it is not the intent of this recommendation to suggest that DOD should concentrate *only* on limited domains. The focused activities recommended here are intended to complement the standards and common infrastructure elements that provide a necessary foundation for mission-specific capabilities.[32]

---

[31] Another candidate mission slice is joint strike—the use of air, sea-based, and land forces to conduct air and artillery strikes behind enemy lines. Execution of this inherently joint mission depends on a demanding set of joint tasks such as dynamic target tasking and managing airspace use (deconfliction).

[32] There are also ongoing efforts by the Air Force Science Advisory Board to explore other broader approaches (in addition to JTA and common infrastructure) that complement the mission slice approach.

Recommendation I-2: The Secretary of Defense should establish a joint C4I integration and interoperability activity to address integration and interoperability throughout the entire life cycle of C4I systems.

The committee believes that an appropriate augmentation of current DOD activities for promoting and facilitating C4I interoperability should focus on three areas: more cross-service testing that starts early in the development process, an increased emphasis on end-to-end field testing, and greater end-to-end interoperability support in operational contexts.

- *Cross-service testing starting early in the development process.* In particular, the testing component of current interoperability efforts is technically oriented and directed primarily at system-to-system testing when the development effort nears completion. Testing in general is oriented toward standards compliance or system acceptance rather than mission performance. The committee believes that to the extent possible, interoperability should be analyzed, assessed, and driven "top-down" by considerations of operational significance as well as facilitated "bottom-up" by the C4I technical community. Focusing on systems within a mission slice, such testing would augment current efforts by testing at application and data layers much earlier in the development process than current practice, perhaps even against paper requirements. Early attention to system-to-system testing for interoperability would make it easier to synchronize the objectives and time lines of different programs for C4I systems that must interoperate, reduce the effort needed to achieve interoperability, and decrease the time line required for addressing interoperability problems in the field by providing "preventive medicine." In addition, it would avoid the cost and complexity entailed when problems are fixed late in the development process. Note that a cross-service development activity is consistent with the desire articulated by the Secretary of Defense in response to Section 912(c) of the Defense Authorization Act for FY 1998. Specifically, the Secretary of Defense expressed interest in ways to establish a joint command, control, and communication integrated system development process that focuses on developing a joint architecture to guide design and achieve integrated systems development.[33]
- *Ongoing interoperability assurance in operational contexts.* Notwithstanding "best efforts" to address interoperability problems before systems are fielded, unanticipated interoperability problems invariably arise as C4I systems are composed and configured in untested combinations in

---

[33]See William S. Cohen. 1998. *Secretary of Defense Report to Congress: Actions to Accelerate the Movement to the New Workforce Vision*, Department of Defense, Washington, D.C.

the field to support particular operational needs. Such problems are increasingly complex, demanding technical sophistication to address "higher-level" issues such as data interoperability. DOD does undertake some activities that address end-to-end C4I interoperability, but the committee is unaware of a process or mechanism in the DOD that systematically addresses C4I interoperability on an end-to-end basis, in a real-world operational setting, in a manner that provides assurance for commanders who will need to support critical operational jobs.

- *Interoperability support for deployed forces.* Deployable operational support for interoperability would help joint task force commanders to address and respond to interoperability problems and issues as they inevitably arise in the field. Deployed in much the same way that the Joint Communications Support Element deploys, "field interoperability support teams" would deploy as an initial cadre with a joint task force first to guide the interconnection of C4I systems in-theater and then to provide sustaining support as required, focusing on integrating and ensuring the interoperability all CINC/service-provided C4I systems (specifically including decision-making/decision-executing information systems as well as communications systems). Field integration/interoperability support teams would involve designated sets of broadly skilled information systems and networking personnel and tools to support their work. Using specialists who are intimately familiar with C4I systems being deployed would increase the likelihood that interoperability problems could be resolved under the extreme time pressures that characterize many operational deployments. Because they would be intimately familiar with the design and development foundations of fielded C4I systems, field integration support teams would have an important advantage over today's deployed "signal" units with respect to certain classes of problems. In addition, they would have the technical background to be able to speak knowledgeably to system suppliers and vendors to obtain high levels of technical support to fix interoperability problems in the field (e.g., by obtaining software patches).[34]

The committee also notes that force interoperability would be more easily achieved, and the burden of field integration reduced, if planning for contingencies were less ad hoc. Deliberate operational planning—

---

[34]Currently, commanders may find it impossible to solve significant interoperability problems between C4I systems rapidly enough to be useful in a contingency. They thus must resort to suboptimal work-arounds employing whatever communications links can be established. With appropriate significant technical support, more problems could be solved rapidly enough to be relevant to an operational deployment.

perhaps in conjunction with predesignation of joint task force commanders—by the force provider would permit much better advance knowledge of which forces would be called upon to interact in a given contingency. With such knowledge, the force provider would be in a better position to focus testing, training, and the like, so as to maximize the interoperability of forces that would be called upon to carry out a particular deployment.

Development, field testing, and operational support draw on the same knowledge and experience base and are inherently synergistic—testing builds expertise that is valuable for field support, and field activities improve future design, development, and test efforts. Also, as an organization that includes both developers and testers, and that has direct contact with end users in the field, a joint C4I integration and interoperability activity would provide a collaborative environment that would foster less adversarial relationships more akin to those increasingly evident in the commercial sector (see section 2.4 and Box 2.8).

To build and sustain the expertise that field integration/interoperability teams need in order to be able to address interoperability problems "on the fly," the joint C4I integration and interoperability activity would have a peacetime role that includes:

• *Interaction with users.* The joint C4I integration and interoperability activity would interact with users in the conduct of acceptance tests as well as the subsequent readiness tests that are used by forces in deciding whether to field an accepted system. During its site visits, the committee observed several instances of such collaboration.[35]

• *Conducting cross-service/agency interoperability tests.* This work would include identifying and synchronizing test opportunities as programs progress through their individual development cycles (within the spiral model framework).

• *Participation in joint and service-sponsored tests and exercises.*

• *Education and training of personnel on interoperability issues.* For example, the joint C4I integration and interoperability activity would train field operational personnel to prevent the decreased interoperability that can result from applying ad hoc solutions to field problems.

---

[35]In its site visits, the committee observed a close interaction between users and developers at the Force XXI Advanced Warfighting Experiments in Fort Hood. It also observed the USAF 605th Test Squadron, which is responsible both for acceptance tests that are part of the acquisition cycle (i.e., tests that the new components satisfy a contractual requirement) and readiness tests that are part of a force's decision to deploy an accepted C4I system (i.e., tests that the force knows how to combine the new C4I component with existing components and how to apply them). Both of these examples illustrate the right synergy between developers and users.

- *C4I configuration and version management.* The joint C4I integration and interoperability activity would develop rules and guidelines to identify configurations of C4I systems—including software and hardware versions—that are known to interoperate with others and provide codified guidance to operational commanders as to what units and C4I systems would interoperate in a deployment.
- *Development of fixes and work-arounds to resolve interoperability problems.* Dealing with interoperability topics at the "higher level" demands being conversant with variations among successive software and hardware releases associated with a particular product line or functionality. This knowledge would also be transferred to operational commands in the form of advice as to what units and systems could and could not be made to interoperate in a deployment.

Doing the work described above would keep personnel familiar with the "building blocks" that could be delivered to the field and from which an end-to-end joint capability would be configured.

During exercises and operational deployments, the joint C4I integration and interoperability activity would offer advice to commanders in planning deployments and provide field support to fix interoperability problems. The technical expertise regarding C4I systems, together with the operational perspective gained from its involvement with joint exercises and deployments, makes the joint C4I integration and interoperability activity a powerful mechanism for improving the coupling between the development community and the community of users.

Where Should the Joint C4I Integration and Interoperability Activity Be Established?

The committee believes that specifying function is more important than specifying organization. Nonetheless, it notes that several organizations—both existing and proposed—have missions that overlap with the proposed joint C4I integration and interoperability activity:

- The Joint Interoperability Test Center in Fort Huachuca, Arizona, plays an important role in technical interoperability testing. Also, while the Joint Interoperability Test Center does not have a formal role for testing interoperability across systems from different services at a functional or operational level (nor does any other DOD agency or organization), its personnel in fact do provide a measure of operational support to exercises. In fact, briefings from the Joint Interoperability Test Center suggested to the committee that this organization is placing increasing emphasis on field support.

- Cross-service development activities are undertaken by the major service C4I developers (the Air Force Electronic Systems Command, the Army Communications and Electronics Command, and the Navy Space and Naval Warfare Systems Command), the associated program executive officers and program managers, the Defense Information Systems Agency, and end users. Existing and emerging service developer capabilities include the Electronic Systems Command's Command and Control Unified Battlespace Environment, the Communication and Electronics Command's Digital Integration Laboratory, and the Space and Naval Warfare Systems Command's Integration Laboratory and Advanced Concepts site, as well as other efforts to establish a joint developer's test bed. Including service developers in an activity that provides field support would add a measure of detailed expertise that is often needed to resolve interoperability problems in the field.
- The proposed interconnection of service/agency development facilities resembles the concept of the battle laboratory "federation" associated with the recently formed Joint Battle Center but is intended to emphasize developmental phase, working level, technically based testing prior to and after formal evaluations.
- The Under Secretary of Defense for Acquisition and Technology and the Assistant Secretary of Defense for C3I were tasked by the Secretary of Defense to examine ways to achieve objectives that are, in part, similar to the proposed joint C4I integration and interoperability activity.[36] The tasking specifically requested submission of an implementation plan to streamline the acquisition organizations, work force, and infrastructure. Despite the general focus on streamlining and infrastructure redirection, the section of the Secretary's report devoted to the restructuring of research, development, testing, and evaluation identified C4I shortfalls vis-à-vis the conduct of joint operations and directed that a study group be formed to address responsive improvements.[37]

---

[36]This tasking derived from an April 1, 1998, Secretary of Defense response to congressional direction in Section 912 (c) of the national Defense Authorization Act for Fiscal Year 1998.

[37]See William S. Cohen. 1998. *Secretary of Defense Report to Congress: Actions to Accelerate the Movement to the New Workforce Vision,* Department of Defense, Washington, D.C. Quoting from the Secretary's report: "From Grenada in 1983 to Operation Desert Storm in 1991, joint operations have been hindered by the inability of forces to share critical information at the rate and at the locations demanded by modern warfare. To attack this problem, I will direct the Under Secretary of Defense (Acquisition & Technology) and the Assistant Secretary of Defense (Command, Control, Communications, and Intelligence) to create a study group that will examine ways to establish a joint command, control and communication integrated system development process that focuses on developing a joint architecture to guide design and achieve integrated systems development."

A study group composed of the commanders of the three primary service C4I acquisition commands (the Air Force Electronic Systems Command, the Army Communications and Electronics Command, and the Navy Space and Naval Warfare Systems Command) was established and, under the guidance of the Under Secretary of Defense for Acquisition and Technology and the Assistant Secretary of Defense for C3I, has evolved a concept for improving C4I integration and interoperability. As of this writing, the fundamental concept involves both (1) working across the acquisition/development commands and related organizations to ensure that systems are "built and tested joint" and (2) establishing tri-service acquisition/development command activities to respond to the needs and problems of the CINCs. An acquisition/development command management forum, the Joint C2 Integration and Interoperability Group, which reports to the Assistant Secretary of Defense for C3I and has mechanisms in place to incorporate the perspectives of other stakeholders, has been established.

The committee understands that a study report and implementation plan have been submitted to the Undersecretary of Defense for Acquisition and Technology. Although the specifics are still evolving, the committee applauds this initiative and views it as potentially addressing the intent of the cross-service development component of the committee's recommendation. As of this writing it is less clear if that would satisfy the on-call, high-level field support component also included in the committee's recommendation.

Recommendation I-3: The Secretary of Defense, the Assistant Secretary of Defense for C3I, and the Chairman of the Joint Chiefs of Staff should establish processes to assess C4I interoperability on a regular basis.

Interoperability is typically assessed by DOD through non-comprehensive perspectives that are focused, for example, on standards, COE compliance, data models, or certification criteria and how individual systems match up to such criteria or standards.

Although a definitive, rigorous analytical approach at a detailed level could be sought, what is key is an approach that provides an overview for management attention at upper levels of both the user (e.g., Joint Chiefs of Staff and the CINCs) and the developer (e.g., the Assistant Secretary of Defense for C3I and the services) communities. In this approach, the interoperability management challenge becomes one of addressing the state of interoperability throughout the enterprise, rather than attempting to strictly measure every variable in detail in each possible scenario.

Recommendation I-3.1: The Assistant Secretary of Defense for C3I and the Joint Chiefs of Staff should develop a set of "interoperability scorecards" as a basis for management, covering the spectrum from compliance with standards to successful end-to-end mission support.

A scorecard approach, which would be useful in assessing the status of cross-service C4I interoperability, is recommended. This approach would support problem prioritization, diagnosis, and correction, as well as operationally based assessment of the state of C4I interoperability for the use of system managers (e.g., the Office of the Assistant Secretary of Defense for C3I and the Military Communications and Electronics Board) and operational users (e.g., CINCs or joint task force commanders). The scorecard approach cannot provide sophistication and quantitative theoretically grounded measures. Rather, the fundamental motivator is to move to a point that interoperability is analyzed, assessed, and driven top-down by considerations of operational significance.

The approach uses three scorecards—technical, systems, and operational—corresponding to the elements of the architectural triad. The technical compliance scorecard would be used to assess a system's implementation from a technical interoperability perspective. It would take the form of a system profile or list for scoring each system implementation with respect to compliance or non-compliance with relevant standards and guidance. The systems interoperability scorecard would measure system-to-system interoperability and would take the form of a matrix displaying the ability of all pairs of systems to interact with each other. The operational interoperability scorecard would be used to assess the ability of a set of systems to satisfy specific node-to-node information flow requirements as well as the collective set of information flows needed to satisfy a defined mission or mission slice.

Assessment requires that responsibility is assigned for (a) the development and definition of criteria, (b) actual conduct of the assessment, as well as (c) responsibility for ensuring that the results of the assessments translate into actions to remedy issues uncovered in the assessment. Responsibility should be assigned as follows:

   a. *Development and definition of criteria.* For assessment of technical and systems interoperability the committee believes that the Assistant Secretary of Defense for C3I should have the lead in development of criteria, and definition of operational interoperability criteria should be a joint responsibility of the Joint Chiefs of Staff and the combat commands.
   b. *Conduct of interoperability assessments.* Given the Defense Information Systems Agency's Joint Interoperability Test Center's role in testing compliance with the Joint Technical Architecture, it is appropriate

that these organizations conduct the technical interoperability assessment. System developers and the organization proposed in Recommendation I-2 above are the logical responsible organizations to assess system interoperability. The operational interoperability assessments should be conducted by an appropriate organization as tasked by the Joint Chiefs of Staff (e.g., the Joint Theater Air Missile Defense Organization for theater missile defense). The responsible organizations should ensure that users play a key role in making such assessments. Without a process for continual assessment for interoperability, the user will have little sense for what interoperability problems need fixing and what their impact might be.

c. *Accountability for the results of the assessments.* Accountability and responsibility for remedying shortcomings uncovered should lie with the Assistant Secretary of Defense for C3I in the case of the technical and system assessments and with the Joint Chiefs of Staff and the operational commands in the case of operational interoperability assessments.

Recommendation I-3.2: The Chairman of the Joint Chiefs of Staff should establish a process to incorporate C4I interoperability into readiness reporting.

Achieving and maintaining an adequate level of interoperability pose significant challenges. Indeed, given the large number of C4I systems that are increasingly expected to interoperate and given the increasing rate at which new hardware and software versions are being fielded, especially as the spiral development model is adopted, the state of interoperability in fielded systems requires ongoing attention.

To help in assessing progress toward meeting these challenges, C4I interoperability should be built into readiness reporting. It must, however, be done in a way that recognizes the characteristics of current combat readiness reporting, which emanates from the combat unit (i.e., a battalion, squadron, or ship) and reports on the status of things controlled directly by the unit or supported directly by its parent command (e.g., training, manning, availability of spare parts, in-commission rates for combat equipment).

The level at which C4I interoperability readiness can be measured and reported differs from the level of the factors included in current combat readiness reporting. Individual combat units are not necessarily in a position to ascertain the status of C4I systems external to their own units—the key issue in determining the ability to conduct joint combat operations. It is at echelons of command above combat units, particularly those with a joint perspective, where there is today no formal combat readiness reporting system, that the readiness of C4I systems can best be assessed.

Also, C4I is a more indirect contributor to combat power than factors that can be directly tied to a particular units' combat readiness.

The system that the Joint Chiefs of Staff develops must focus on assessing the readiness of forces to conduct end-to-end missions. The readiness reporting must be based on a realistic set of scenarios for how units are to be employed. It may be appropriate to focus assessment efforts in the same mission slices as those in which the activities proposed in Recommendation I-1 are conducted.

The joint force provider, the U.S. Atlantic Command, is in a unique position to provide cross-service visibility of the contribution of interoperability factors to readiness. Its ability to do so would be enhanced if it were to preplan which forces would be expected to be deployed in a given contingency, an approach that would also enable better assessment of interoperability as an element of readiness for that contingency.

Recommendation I-4: The services and agencies should designate an activity within the program offices for C4I systems (and weapons systems with embedded C4I) to be explicitly responsible for resolution of architectural and system-level issues that determine interoperability.

The committee believes that management attention to the interoperability of C4I systems is today often not an assured process. That is, while today's acquisition system does acknowledge the importance of C4I interoperability in its various directives, in actual practice the management attention afforded to C4I interoperability depends strongly on the temperament and inclinations of the individual program manager. Program managers who "buy into" the vision of interoperability pay a great deal of attention to that subject, while those who do not pay far less attention to it. The focus of this recommendation is not the development of additional acquisition policy but rather the filling of a gap in implementation and practice in program management.

The establishment of an "interoperability cell" or equivalent in C4I program offices would provide a central point of focus for interoperability issues. The fundamental principle of this activity is that for each C4I program there must be an activity charged with looking outward, cross-service, at interoperability issues. This cell would be responsible for revising or modifying architecture as needed to accommodate changes in doctrine, tactics, techniques, procedures, and equipment; engaging the stakeholders in a particular C4I system in the problem of defining and achieving interoperability for that system; and negotiating interoperability issues with those responsible for development/upgrade of "neighboring" systems. In its efforts, an interoperability cell could be expected to take a pragmatic approach that narrows the scope of the systems that must be considered by the architectures (though not necessarily to those that are immediately adjacent to any given program), and limit the scope of the interoperability effort to particular configurations of components. In other words, interoperability cannot be realistically expected across an unlimited number of releases or versions. Such "bottom-up" negotiation of interoperability issues is intended to complement "top-down" architectural and common infrastructure efforts. The cell would also be responsible for elements that constitute an investment in future interoperability, such as metadata and appropriate interfaces.

The appropriate placement for an interoperability activity can vary. For a large program, placement within the program manager's office may be advantageous. In other cases, placement within a program executive office or service acquisition command may make sense—both for reasons of efficiency and to ensure that the "cell" has a sufficiently broad perspective. Somewhat similar functions have been performed by service elements such as the Air Force Electronic Systems Command, the Navy Space and Naval Warfare Systems Command, and the Army Communications and Electronics Command. However, there is no clear analogy to these organizations for joint systems.

3

Information Systems Security

3.1 INTRODUCTION

DOD's increasing reliance on information technology in military operations increases the value of DOD's information infrastructure and information systems as a military target. Thus, for the United States to realize the benefits of increased use of C4I in the face of a clever and determined opponent, it must secure its C4I systems against attack.

As noted in Chapter 2, the maximum benefit of C4I systems is derived from their interoperability and integration. That is, to operate effectively, C4I systems must be interconnected so that they can function as part of a larger "system of systems." These electronic interconnections multiply many-fold the opportunities for an adversary to attack them.

Maintaining the security of C4I systems is a problem with two dimensions. The first dimension is physical, that of protecting the computers and communications links as well as command and control facilities from being physically destroyed or jammed. For this task, the military has a great deal of relevant experience that it applies to systems in the field. Thus, the military knows to place key C4I nodes in well-protected areas, to put guards and other access control mechanisms in place to prevent sabotage, and so on. The military also knows how to design and use wireless communications links so that enemy jamming is less of a threat.

Information systems security is a much more challenging task. Information systems security—the task of protecting the C4I systems connected to the communications network against an adversary's information attack against those systems—is a much more poorly understood area than physical security.[1] Indeed, DOD systems are regularly attacked and penetrated,[2] though most of these attacks fail to do damage. Recent exercises such as Eligible Receiver (Box 3.1) have demonstrated real and significant vulnerabilities in DOD C4I systems, calling into question their ability to perform properly when faced with a serious attack by a determined and skilled adversary.

Such observations are unfortunately not new. A series of earlier reports have noted a history of insufficient or ineffective attention to C4I information systems security (Box 3.2).

The problem of protecting DOD C4I systems against attack is enormously complicated by the fact that DOD C4I systems and the networks to which they are connected are not independent of the U.S. national information infrastructure.[3] Indeed, the line between the two is quite blurred because many military systems make use of the civilian information infrastructure,[4] and because military and civilian systems are often interconnected. DOD is thus faced with the problem of relying on components of the infrastructure over which it does not have control. While the general principles of protecting networks as described below apply to military C4I systems, both those connected to civilian components and those that are not, the policy issues related to DOD reliance on the national information infrastructure are not addressed in this report. Lastly, C4I systems are increasingly built upon commercial technologies and thus

---

[1]Within the information technology industry, the term "information security" encompasses technical and procedural measures providing for confidentiality, authentication, data integrity, and non-repudiation, as well as for resistance to denial-of-service attacks. The committee understands that within many parts of DOD, the term "information security" does not have such broad connotations. Nevertheless, it believes that lack of a broad interpretation for the term creates problems for DOD because it focuses DOD on too narrow a set of issues. Note that information systems security does *not* address issues related to the quality of data before it is entered into the C4I system. Obviously, such issues are important to the achievement of information superiority, but they are not the focus of this chapter.

[2]In 1996, the General Accounting Office reported that the DOD may have experienced 250,000 cyber-attacks in 1995 and that the number of cyber-attacks would increase in the future. Furthermore, the Defense Information Systems Agency estimated that "only about 1 in 50 attacks is actually detected and reported." For additional information, see General Accounting Office. 1996. *Information Security: Computer Attacks at the Department of Defense Pose Increasing Risks,* GAO/AIMD-96-84, General Accounting Office, Washington, D.C.

[3]The U.S. national information infrastructure includes those information systems and networks that are used for all purposes, both military and civilian, whereas DOD's C4I systems are by definition used for military purposes.

[4]More than 95 percent of U.S. military and intelligence community voice and data communications are carried over facilities owned by public carriers. (See Joint Security Commission, *Redefining Security: A Report to the Secretary of Defense and the Director of Central Intelligence,* February 28, 1994, Chapter 8.)

BOX 3.1 Eligible Receiver

Conducted in the summer of 1997 and directed by the Chairman of the Joint Chiefs of Staff, Eligible Receiver 97 was the first large-scale no-notice DOD exercise (a real, not tabletop, exercise) designed to test the ability of the United States to respond to an attack on the DOD and U.S. national infrastructure. This exercise involved a simulated attack against components of the national infrastructure (e.g., power and communications systems) and an actual "red team" attack against key defense information systems at the Pentagon, defense support agencies, and in combatant commands.

The attack on the national infrastructure was based on potential vulnerabilities, while the actual attack on defense systems exploited both actual and potential vulnerabilities. (The vulnerabilities exploited were common ones, including bad or easily guessed passwords, operating system deficiencies, and improper system configuration control, sensitive site-related information posted on open Web pages, inadequate user awareness of operational security, and poor operator training.) All red team attacks were based on information and techniques derived from open non-classified research, and no insider information was provided to the red team. Furthermore, the red team conducted extensive "electronic reconnaissance" before it executed its attacks.

The exercise demonstrated a high degree of interdependence between the defense and national information infrastructures. For example, the defense information infrastructure is extremely reliant on commercial computer and communication networks, and the public and private sectors often share common commercial software or systems. As a result, vulnerabilities demonstrated in DOD systems and procedures may be shared by others, and vulnerabilities in one area may allow exploitation in other areas.

The exercise revealed vulnerabilities in DOD information systems and deficiencies in the ability of the United States to respond effectively to a coordinated attack on the national infrastructure and information systems. Poor operations and information security practices provided many red team opportunities. In short, the exercise provided real evidence of network vulnerabilities.

BOX 3.2 Some Related Studies on Information Security

Computers at Risk: Safe Computing in the Information Age[1] focused on approaches for "raising the bar" of computer and communications security so that all users—both civilian and military—would benefit, rather than just those who are users and handlers of classified government information. The report responded to prevailing conditions of limited awareness by the public, system developers, system operators, and policymakers. To help set and raise expectations about system security, the study recommended:

- Development and promulgation of a comprehensive set of generally accepted security system principles;
- Creation of a repository of data about incidents;
- Education in practice, ethics, and engineering of secure systems; and
- Establishment of a new institution to implement these recommendations.

Computers at Risk also analyzed and suggested remedies for the failure of the marketplace to substantially increase the supply of security technology; export control criteria and procedures were named as one of many contributing factors. Observing that university-based research in computer security was at a "dangerously low level," the report mentioned broad areas where research should be pursued.

The 1996 *Report of the Defense Science Board Task Force on Information Warfare Defense*[2] focused on defending against cyber-threats and information warfare. The task force documented an increasing military dependence on networked information infrastructures, analyzed vulnerabilities of the current networked information infrastructure, discussed actual attacks on that infrastructure, and formulated a list of threats that has been discussed broadly within the DOD and elsewhere. The task force concluded that "there is a need for extraordinary action to deal with the present and emerging challenges of defending against possible information warfare attacks on facilities, information, information systems, and networks of the United States which [sic] would seriously affect the ability of the Department of Defense to carry out its assigned missions and functions."

Some of the task force recommendations answered organizational questions, e.g., where within DOD various information warfare defense functions might be placed, how to educate senior government and industry leaders about vulnerabilities and their implications, and how to determine current infrastructure dependencies and vulnerabilities. Other recommendations addressed short- and longer-term technical means for repelling attacks. The task force urged greater use of existing security technology, certain controversial encryption technology, and the construction of a minimum essential information infrastructure. The task force noted the low levels of activity concerning computer security and survivable systems at universities, and also suggested a research program for furthering the development of the following:

continues

- System architectures that degrade gracefully and are resilient to failures or attacks directed at single components;
- Methods for modeling, monitoring, and managing large-scale distributed systems; and
- Tools and techniques for automated detection and analysis of localized or coordinated large-scale attacks, and tools and methods for predicting anticipated performance of survivable distributed systems.

Trust in Cyberspace[3] proposed a research agenda for building networked systems that are more robust, reducing software design problems, and developing mechanisms to protect against new types of attacks from unauthorized users, criminals, or terrorists. The report noted that much of today's security technology for operating systems is based on a model of computing centered on mainframe computers. Today, different security mechanisms are needed to protect against the new classes of attacks that become possible because of computer networks, the distribution of software using the Internet, and the significant use of commercial, off-the-shelf (COTS) software. Furthermore, the report recommended a more pragmatic approach to security that incorporates add-on technologies, such as firewalls, and utilizes the concept of defense in depth, which requires independent mechanisms to isolate failures so that they do not cascade from one area of the system to another.

In the area of network design, the report noted a need for research to better understand how networked information systems operate, how their components work together, and how changes occur over time. Since a typical computer network is large and complex, few engineers are likely to understand the entire system. Better conceptual models of such systems will help operators grasp the structure of these networks and better understand the effects of actions they may take to fix problems. Approaches to designing secure networks built from commercially available software warrant attention. Improvements in testing techniques and other methods for determining errors also are likely to have considerable payoffs for enhancing assurance in networked systems.

Finally, research is needed to deal with the major challenges for network software developers that arise because COTS components are used in the creation of most networked information systems. Indeed, today's networked information systems must be developed with limited access to significant pieces of the system and virtually no knowledge of how those pieces were developed.

---

[1]Computer Science and Telecommunications Board, National Research Council. 1991. *Computers at Risk: Safe Computing in the Information Age,* National Academy Press, Washington, D.C.

[2]Defense Science Board. 1996. *Report of the Defense Science Board Task Force on Information Warfare-Defense (IW-D),* Office of the Under Secretary of Defense for Acquisition and Technology, Washington, D.C.

[3]Computer Science and Telecommunications Board, National Research Council. 1999. *Trust in Cyberspace,* National Academy Press, Washington, D.C.

are coming to suffer from the same basic set of vulnerabilities that are observed in the commercial sector.

3.1.1 Vulnerabilities in Information Systems and Networks[5]

Information systems and networks can be subject to four generic vulnerabilities. The first is *unauthorized access to data*. By surreptitiously obtaining sensitive data (whether classified or unclassified) or by browsing a sensitive file stored on a C4I computer, an adversary might obtain information that could be used against the national security interests of the United States. Moreover, even more damage could occur if the fact of unauthorized access to data were to go unnoticed, because it would be impossible to take remedial action.

The second generic vulnerability is *clandestine alteration of data*. By altering data clandestinely, an adversary could destroy the confidence of a military planner or disrupt the execution of a plan. For example, alteration of logistics information could significantly disrupt deployments if troops or supplies were rerouted to the wrong destinations or supply requests were deleted.

A third generic vulnerability is *identity fraud*. By illicitly posing as a legitimate user, an adversary could issue false orders, make unauthorized commitments to military commanders seeking resources, or alter the situational awareness databases to his advantage. For example, an adversary who obtained access to military payroll processing systems could have a profound effect on military morale. An enemy who overruns a friendly position and gains access to the information network of friendly forces may see classified information with tactical significance or be able to insert bad information into friendly tactical databases.

A fourth generic vulnerability is *denial of service*. By denying or delaying access to electronic services, an adversary could compromise operational planning and execution, especially for time-critical tasks. For example, attacks that resulted in the unavailability of weather information systems could delay planning for military operations. Attacks that deny friendly forces the use of the Global Positioning System (e.g., through jamming) could cripple targeting of hostile forces and prevent friendly forces from knowing where they are. Denial of service is, in the view of many, the most serious vulnerability, because denial-of-service attacks are relatively easy to carry out and often require relatively little technical sophistication.

---

[5]Adapted from Computer Science and Telecommunications Board, National Research Council. 1996. *Cryptography's Role in Securing the Information Society*, National Academy Press, Washington, D.C., Box 1.3.

Also, it is worth noting that many compromises of security result not from a successful direct attack on a particular security feature intended to guard against one of these vulnerabilities, but instead from the "legitimate" use of designed-in features in ways that were not initially anticipated by the designers of that feature. Thus, defense must be approached on a system level rather than on a piecemeal basis.

Lastly, non-technical vulnerabilities—such as the intentional misuse of privileges by authorized users—must be considered. For example, even perfect access controls and unbreakable encryption will not prevent a trusted insider from revealing the contents of a classified memorandum to unauthorized parties.

The types of attack faced by DOD C4I systems are much broader and potentially much more serious and intense than those usually faced by commercial (non-military) networked information systems. The reason is that attacks on DOD C4I systems that are part of an attack sponsored or instigated by a foreign government can draw upon virtually unlimited resources devoted to those attacks. Furthermore, perpetrators sponsored or supported by a foreign government are largely immune to retaliation or punishment through law enforcement channels, and are thus free to act virtually without constraint.

3.1.2 Security Requirements

Needs for information systems security and trust can be formulated in terms of several major requirements:

- **Data confidentiality**—controlling who gets to read information in order to keep sensitive information from being disclosed to unauthorized recipients, e.g., by preventing the disclosure of classified information to an adversary;
- **Data integrity**—assuring that information and programs are changed, altered, or modified only in a specified and authorized manner, e.g., by preventing an adversary from modifying orders given to combat units so as to shape battlefield events to his advantage;
- **System availability**—assuring that authorized users have continued and timely access to information and resources, e.g., by preventing an adversary from flooding a network with bogus traffic that delays legitimate traffic such as that containing new orders from being transmitted; and
- **System configuration**—assuring that the configuration of a system or a network is changed only in accordance with established security guidelines and only by authorized users, e.g., by detecting and reporting to higher authority the improper installation of a modem that can be used for remote access.

In addition, there is a requirement that cuts across these four, the requirement for **accountability**—knowing who has had access to information or resources.

It is apparent from this listing that security means more than protecting information from disclosure (e.g., classified information). In the DOD context, much of the information on which military operations depend (e.g., data related to personnel, payroll, logistics, and weather) is not classified. While its *disclosure* might not harm national security, alteration or a delay in transmitting it certainly could.[6] In other cases, access to unclassified information can present a threat (e.g., access to personnel medical records used to enable blackmail attempts).

Satisfying these security requirements requires a range of security services, including:

- **Authentication**—ascertaining that the identity claimed by a party is indeed the identity of that party. Authentication is generally based on what a party knows (e.g., a password), what a party has (e.g., a hardware computer-readable token), or what a party is (e.g., a fingerprint);
- **Authorization**—granting of permission to a party to perform a given action (or set of actions);
- **Auditing**—recording each operation that is invoked along with the identity of the subject performing it and the object acted upon (as well as later examining these records); and
- **Non-repudiation**—the use of a digital signature procedure affirming both the integrity of a given message and the identity of its creator to protect against a subsequent attempt to deny authenticity.

3.1.3 Role of Cryptography

It is important to understand what role the tool of cryptography plays in information system security, and what aspects of security are not provided by cryptography. Cryptography provides a number of useful capabilities:

- **Confidentiality**—the characteristic that information is protected from disclosure, in transit during communications (so-called link encryp-

---

[6]Statements typically issued by DOD in the aftermath of an identified attack on its systems assure Congress and the public that "no classified information was disclosed." These may be technically correct, but they do not address the important questions of whether military capabilities were compromised, or more broadly, if a similar incident would have adverse implications in future, purposeful attack situations.

tion) and/or when stored in an information system. The security requirement of confidentiality is the one most directly met by cryptography;

- **Authentication**—cryptographically based assurance that an asserted identity is valid for a given person or computer system;
- **Integrity check**—cryptographically based assurance that a message or file has not been tampered with or altered; and
- **Digital signature**—assurance that a message or file was sent or created by a given person, based on the capabilities provided by mechanisms for authentication and integrity checks.

Cryptographic devices are important, for they can protect information in transit against unauthorized disclosure, but this is only a piece of the information systems security problem. The DOD mission also requires that information be protected while in storage and while being processed, and that the information be protected not only against unauthorized disclosure, but also against unauthorized modification and against attacks that seek to deny authorized users timely access to the information.

Cryptography is a valuable tool for authentication as well as for verifying the integrity of information or programs.[7] Cryptography alone does not provide availability (though because its use is fundamental to many information security measures, its widespread application can contribute to greater assurance of availability[8]). Nor does cryptography directly provide auditing services, though it can serve a useful role in authenticating the users whose actions are logged and in verifying the integrity of audit records.

Cryptography does not address vulnerabilities due to faults in a system, including configuration bugs and bugs in cryptographic programs. It does not address the many vulnerabilities in operating systems and applications.[9] It certainly does not provide a solution to such problems as

---

[7]Cryptography can be used to generate digital signatures of messages, enabling the recipient of a message to assure himself that the message has not been altered (i.e., an after-the-fact check of message integrity that does not protect against modification itself). However, in the larger view, a properly encrypted communications channel is difficult to compromise in the first place, and in that sense cryptography can also help to prevent (rather than just to detect) improper modifications of messages.

[8]Widespread use of encryption (vs. cryptography) can also result in reduced availability, as it hinders existing fault isolation and monitoring techniques. It is for this reason that today's network managers are often not enthusiastic about deployment of encryption.

[9]Recent analysis of advisories issued by the Computer Emergency Response Team at Carnegie Mellon University indicates that 85 percent of them would not have been solved by encryption. See Computer Science and Telecommunications Board, National Research Council. 1999. *Trust in Cyberspace*, National Academy Press, Washington, D.C.

poor management and operational procedures or dishonest or suborned personnel.

In summary, cryptography may well be a necessary component of these latter protections, but cryptography alone is not sufficient.[10]

3.2 MAJOR CHALLENGES TO INFORMATION SYSTEMS SECURITY

3.2.1 The Asymmetry Between Defense and Offense

Information systems security is fundamentally a defensive function, and as such suffers from an inherent asymmetry between cyber-attack and cyber-defense. Because cyber-attack can be conducted at the discretion of the attacker, while the defender must always be on guard, cyber-attack is often cheaper than defense, a point illustrated by the modest resources used by hackers to break into many unclassified DOD systems. Furthermore, for the defender to be realistically confident that his systems are secure, he must devote an enormous amount of effort to eliminate all security flaws that an attacker might exploit, while the attacker simply needs to find one overlooked flaw. Finally, defensive measures must be developed and deployed, a process that takes time, while attackers generally exploit existing security holes. In short, a successful defender must be successful against all attacks, regardless of where the attack occurs, the modality of the attack, or the time of the attack. A successful attacker has only to succeed in one place at one time with one technique. It is this asymmetry that underlies the threat-countermeasure cycle. A countermeasure is developed and deployed against a known threat, which prompts the would-be attacker to develop another threat. As a result, the advantage is heavily to the attacker until most potential vulnerabilities have been addressed (i.e., after many iterations of the cycle).[11]

3.2.2 Networked Systems

The utility of an information or C4I system generally increases as the number of other systems to which it is connected increases. On the other

---

[10]It is worth noting that cryptography is often the *source* of failures of C4I systems to interoperate. That is, two C4I systems often fail to exchange data operating in secure encrypted mode.

[11]This asymmetry is discussed in Computer Science and Telecommunications Board, National Research Council. 1990. *Computers at Risk: Safe Computing in the Information Age,* National Academy Press, Washington, D.C.

hand, increasing the number of connections of a system to other systems also increases its vulnerability to attacks routed through those connections.

The use of the Internet to connect C4I systems poses special vulnerabilities. It is desirable to use the Internet because the Internet provides lower information transport costs compared to the public switched telephone network or dedicated systems. But the Internet provides neither quality-of-service guarantees nor good isolation from potentially hostile parties.

3.2.3 Ease-of-Use Compromises

Compromises arise because information systems security measures ideally make a system impossible to use by someone who is not authorized to use it, whereas considerations of system functionality require that the system be easy to use by authorized users. From the perspective of an authorized user, a system with information systems security features should look like the same system without those features. In other words, security features provide no direct functional benefit to the authorized user. At the same time, measures taken to increase the information security of a system almost always make using that system more difficult or cumbersome. The result in practice is that all too often (from a security standpoint) security features are simply omitted (or not turned on) to preserve the ease-of-use goal.

3.2.4 Perimeter Defense

Today's commercially available operating systems and networks offer only weak defensive mechanisms, and thus the components that make up a system are both vulnerable and hard to protect. One approach to protecting a network is then to allow systems on the network to communicate freely (i.e., without the benefit of security mechanisms protecting each individual network transaction) while allowing connection to the larger world outside the network only through carefully defined and well-protected gateways. The result is an arrangement that is "hard on the outside" against attack but "soft on the inside." Thus, it is today very common to see "enclaves" hiding from the Internet behind firewalls, but few defensive measures within the enclaves themselves.

A perimeter strategy is less expensive than an approach in which every system on a network is protected (a defense-in-depth strategy) because defensive efforts can be concentrated on just a few nodes (the gateways). But the major risk is that a single success in penetrating the perimeter compromises everything on the inside. Once the perimeter is breached, the attacker need not expend additional effort to increase the number of targets that he may attack. The problem of perimeter defense is made worse by the tendency to let one's guard down within the protection of the firewall (believing that the inside is secure) and thus to not take full advantage of even the (relatively weak) protections afforded by the security built into the network components. The limitations of a perimeter defense are issues that should be redressed by C4I architecture—the paradigm of perimeter defense is an implicit element of today's C4I architecture that needs to be made explicit and changed.

One alternative to perimeter defenses is defense in depth, a strategy that requires an adversary to penetrate multiple independently vulnerable obstacles to have access to all of his targets. The property of "independent vulnerabilities" is key; if the different mechanisms of defense share common-mode vulnerabilities (e.g., all use an operating system with easily exploited vulnerabilities), even multiple mechanisms of defense will be easily compromised. When the mechanisms are independently vulnerable and deployed, the number of accessible targets becomes a strong function of the effort expended by the attacker.

3.2.5 The Use of COTS Components[12]

For reasons of economy, time to completion, and interoperability, networked information systems, including many DOD C4I systems, are increasingly built out of commercial off-the-shelf (COTS) components. But the use of COTS components, especially COTS software (including operating systems, network management packages, e-mail programs, Web browsers, and word processors, among others), can lead to security problems for a number of reasons:

• Increasing functionality and decreasing time to market characterize the COTS software market today—often at the expense of security. The reason is simple—security features and functionality do not usually play a large role in buyer decisions.
• The increased functionality of COTS software is generally associated with high complexity and a large number of bugs. The high complexity means that specifications for COTS components are likely to be incomplete and consequently, system architects may be unaware of some of the vulnerabilities in the building-block components.

---

[12]The discussion in this section is based largely on *Trust in Cyberspace*; see Computer Science and Telecommunications Board, National Research Council. 1999. *Trust in Cyberspace*, National Academy Press, Washington, D.C.

• The developers of COTS software rely on customer feedback as a significant, or even primary, quality assurance mechanism, which can lead to uneven quality levels within the different subsystems or functionality in a COTS product. Even worse, security problems in COTS products may not even be known to the customer.

• The use of COTS components implies a dependence on the vendor for decisions about the component's evolution and the engineering processes used in its construction (notably regarding security). Similarly, the security mechanisms available in a COTS product, if any are present at all, are dictated by the developers of COTS products. Because COTS software is developed for a range of application domains, its security mechanisms are usually not tailored to the specific needs of any particular application area.

• The growing use of COTS components, from a small set of vendors, throughout all segments of the information technology industry suggests a continuing decrease in heterogeneity in the coming years. Thus, the similarity intrinsic in the component systems of a homogeneous collection implies that these systems will share vulnerabilities. A successful attack on one system is then likely to succeed on other systems as well.

• COTS components are often bundled together, and some of the components may be insecure. For example, a given operating system may be bundled by the vendor with a particular authentication package. Even if that authentication package is inadequate, the user may be faced with a choice of abandoning the operating system or using inadequate authentication because of the difficulty of replacing that package.

These factors do not argue that COTS components should not be used to develop networked information systems, but rather that such use should be undertaken with care. For example, wise developers learn to avoid the more complex features of COTS software, because these are the most likely to exhibit surprising behavior and such behavior is least likely to remain stable across releases. When these features cannot be avoided, encapsulating components with wrappers, effectively narrowing their interfaces, can protect against some undesirable behaviors.

Still, in an overall sense, the relationship between the use of COTS software and system security is unclear. Research is needed to improve understanding of this relationship, and of how to use COTS components to build secure systems.

3.2.6 Threats Posed by Insiders

Insiders are those authorized to access some part or parts of a network. When security depends on the defenses built into the perimeter, the coercion or subornation of a single individual on the inside leaves the entire network open to attack to the extent that internal protections are lacking. Controlling the privileges of authorized individuals more finely (i.e., enabling such an individual to use some system resources or capabilities but not others) is only a partial solution, because abuse of the enabled resources is possible.

3.2.7 Passive Defense

Legal and technical constraints preclude retaliation against the perpetrator of an information systems attack (a cyber-attack). Thus, the attacker pays no penalty for failed attacks. He or she can therefore continue attacking unpunished until he or she succeeds or quits.

The following example from physical space illustrates the futility of passive defense. Imagine a situation in which truck bombers in a red truck attempt entry to a military base. The bomb is discovered and they are turned away at the front gate of a military base, but allowed to go away in peace to refine their attack. They return later that day with a bomb in a yellow truck, are again turned away, and again go away in peace to refine their attack. They return still later with a stolen military truck. This time the bomb is undetected, they penetrate the defenses, and they succeed in their attack. A base commander taking this approach to security would be justly criticized and held accountable for the penetration.

Yet in cyberspace passive defense is standard operating procedure. For example, an attacker can use an automatic telephone dialer to dial every number on a military post's telephone exchange looking for modem tones. In a phone probe looking for modem tones, all 10,000 phone numbers may be tested. No sane commander would allow a truck bomber 10,000 unchallenged, penalty-free attempts to drive onto a base. But the same commander today is constrained to routinely allow 10,000 unchallenged, penalty-free attempts to find modems attached to base systems.

None of this is to argue that going beyond passive defense is easy or even appropriate. For example, it is often difficult to identify the actual source of a cyber-attack (as opposed to the most immediate node through which that attack is being prosecuted). A cyber-attacker might well use the computer of some legitimate organization to launch an attack, and retaliation against that computer might well damage it. The opportunities for misleading defense mechanisms or defenders, causing them to retaliate against the wrong source, are numerous. Furthermore, in an international context, retaliation against a foreign nation from which an attack is being routed might be regarded as an act of war. For reasons such as these, passive defense in cyberspace represents both the tradition and the standard operating practice. But over the long run, it is a losing proposition, and inadequate for protection of military operations in cyberspace.

3.3 DEFENSIVE FUNCTIONS

Effective information systems security is based on a number of functions described below. This list of functions is not complete; nevertheless, evidence that all these functions are being performed in an effective and coordinated fashion will be evidence that information systems security is being taken seriously and conducted effectively.

Some of these functions were also noted in the military context by the Defense Science Board, and some by the President's Commission on Critical Infrastructure Protection in its report.[13] These functions are listed here because they are important, and because the committee believes that they have not yet been addressed by the DOD in an effective fashion (as described in the committee's findings below).

Function 1. Collect, analyze, and disseminate strategic intelligence about threats to systems.

Any good defense attempts to learn as much as possible about the threats that it may face, both the tools that an adversary may use and the identity and motivations of likely attackers. In the information systems security world, it is difficult to collect information about attackers (though such intelligence information should be sought). It is, however, much easier to collect and analyze information on technical and procedural vulnerabilities, to characterize both the nature of these vulnerabilities and their frequency at different installations. Dissemination of information about these vulnerabilities enables administrators of the information systems that may be affected to take remedial action.

Function 2. Monitor indications and warnings.

All defenses—physical and cyber—rely to some extent on indications and warning of impending attack. The reason is that if it is known that attack is impending, the defense can take actions to reduce its vulnerability and to increase the effectiveness of its response. This function calls for:

---

[13]See the President's Commission on Critical Infrastructure Protection. 1997. *Critical Foundations: Protecting America's Infrastructures*, Government Printing Office, Washington, D.C. Also, the Defense Science Board. 1996. *Report of the Defense Science Board Task Force on Information Warfare-Defense (IW-D)*, Office of the Under Secretary of Defense for Acquisition and Technology, Washington, D.C.

- *Monitoring of threat indicators.* For example, near-simultaneous penetration attempts on hundreds of military information systems might reasonably be considered an indication of an orchestrated attack. Mobilization of a foreign nation's key personnel known to have responsibility for information attacks might be another indicator. The notion of an "information condition" or INFOCON, analogous to the defense condition (DEFCON) indicator, would be a useful summary device to indicate to commanders the state of the cyber-threat at any given time (Box 3.3). This concept is being developed by various DOD elements but is yet immature.
- *Assessment and characterization of the information attack (if any).* Knowledge of the techniques used in an attack on one information system may facilitate a judgment of the seriousness of the attack. For example, an attack that involves techniques that are not widely known may indicate that the perpetrators have a high degree of technical sophistication.[14]
- *Dissemination of information about the target(s) of threat.* Knowledge of the techniques used in an attack on one information system may enable administrators responsible for other systems to take preventive actions tailored to that type of attack. This is true even if the first attack is unsuccessful, because security features that may have thwarted the first attack may not necessarily be installed or operational on other systems.

Note that dissemination of information about attacks and their targets is required on two distinct time scales. The first time scale is seconds or minutes after the attack is known; such knowledge enables operators of other systems not (yet) under attack to take immediate preventive action (such as severing some network connections). In this instance, alternative means of secure communication may be necessary to disseminate

---

[14] Detection of cyber-attacks can be broadly classified into two categories. The first is known as penetration detection, which is usually based on descriptions of these attacks (if such and such conditions are observed, the system is or has come under attack) or on models that abstract characteristics of a known attack (and can thus detect some different variants—some that were previously unknown—of a known penetration). The second is known as anomaly detection, and is based on the detection of events that are not "normal," i.e., not usual in the context of the monitored system. Anomaly detection tends to generate many false positives (because an anomalous event may in fact reflect something that a legitimate user has never done before rather than the sign of a hostile attack), but it is the only known approach to detecting attacks that were previously unknown. Finally, detection of coordinated penetration attempts on a network is necessary (but not sufficient) to characterize a large-scale attack (i.e., one mounted to challenge the United States as a national entity). Comprehensive attack detection is based on security components that deal with all of these dimensions of an attack, interacting with each other to provide the necessary detection components.

> **BOX 3.3 Information Conditions**
>
> One implementation of information conditions (INFOCONs) is defined by the U.S. Strategic Command (STRATCOM). Beginning with the Defense Science Board report of 1996 identifying a need for structured response to attacks on the nation's information infrastructure,[1] the Information Assurance Division of STRATCOM drafted operating instructions that became the INFOCON program. INFOCONs "provide a set of pre-established measures to assess threats against STRATCOM's information systems and define graduated actions to be taken in response to those threats." On a day-to-day basis, the INFOCON is set at "normal," and only routine security measures are taken. If increased hostile actions are detected, INFOCONs are increased to raise information assurance awareness, with higher INFOCONs representing more intense hostile activity and more rigorous response actions.
>
> INFOCONs are roughly analogous to defense condition (DEFCON) and terrorist condition (THREATCON) levels. The decision to change the INFOCON is based on the assessed threat, the capability to implement the required protective measures, and the overall impact the action will have on STRATCOM's capability to perform its mission. INFOCONs define appropriate information operations measures to be taken. Each INFOCON is designed to produce detection, assessment, and response measures commensurate with the existing threat. Escalating INFOCONs enhance information operations capabilities and send a clear signal of increased readiness. Different INFOCONs are not necessarily linear in nature as an organized malicious information attack could immediately require higher INFOCONs to be set and appropriate measures taken.
>
> INFOCON procedures received their first full-scale workout during STRATCOM's annual readiness exercise Global Guardian 98. STRATCOM officials believe that exercise results demonstrated the ability of INFOCONs to raise security awareness and to counter hostile actions. For example, based on independent monitoring of communications during Global Guardian, STRATCOM officials believe that improved operations security practices were demonstrated as compared to previous exercises—an improvement attributed in part to the new INFOCONs.
>
> ---
>
> SOURCE: Adapted from Charles A. Keene, U.S. Strategic Command, "INFOCONs Increase Focus on Information Security," and (no author) "USSTRATCOM Information Operations Conditions," *intercom on-line*, January 1998, Vol. 4, No. 1; available online at <http://www.afca.scott.af.mil/pa/public/98jan/intercom.htm>.
>
> [1]Defense Science Board. 1996. *Report of the Defense Science Board Task Force on Information Warfare-Defense (IW-D)*, Office of the Under Secretary of Defense for Acquisition and Technology, Washington, D.C.

such information. The second time scale is days after the attack is understood; such knowledge allows operators throughout the entire system of systems to implement fixes and patches that they may not yet have fixed, and to request fixes that are needed but not yet developed.

A DOD example of monitoring is the Air Force Information Warfare Center element that monitors and responds to penetrations at Air Force installations worldwide from San Antonio, Texas.

Function 3. Be able to identify intruders.

Electronic intruders into a system are admittedly hard to identify. Attacks are conducted remotely, and a chain of linkages from the attacker's system to an intermediate node to another and another to the attacked system can easily obscure the identity of the intruder. Nevertheless, certain types of information—if collected—may shed some light on the intruder's identity. For example, some attackers may preferentially use certain tools or techniques (e.g., the same dictionary to test for passwords), or use certain sites to gain access. Attacks that go on over an extended period of time may provide further opportunities to trace the origin of the attack.

Function 4. Test for security weaknesses in fielded and operational systems.

An essential part of a security program is searching for technical and operational or procedural vulnerabilities. Ongoing tests (conducted by groups often known as "red teams" or "tiger teams") are essential for several reasons:

- Recognized vulnerabilities are not always corrected, and known fixes are frequently found not to have been applied as a result of poor configuration management.
- Security features are often turned off in an effort to improve operational efficiency. Such actions may improve operational efficiency, but at the potentially high cost of compromising security, sometimes with the primary damage occurring in some distant part of the system.
- Some security measures rely on procedural measures and thus depend on proper training and ongoing vigilance on the part of commanders and system managers.
- Security flaws that are not apparent to the defender undergoing an inspection may be uncovered by a committed attacker (as they would be uncovered in an actual attack).

Thus, it is essential to use available tools and conduct red team or tiger team probes often and without warning to test security defenses. In order to maximize the impact of these tests, reports should be disseminated widely. Release of such information may risk embarrassment of certain parties or possible release of information that can be used by adversaries to attack, but especially in the case of vulnerabilities uncovered for which fixes are available, the benefits of releasing such information—allowing others to learn from it and motivating fixes to be installed—outweigh these costs.[15]

Tiger team attacks launched without the knowledge of the attacked systems also allow estimates to be made of the frequency of attacks. Specifically, the fraction of tiger team attacks that are detected is a reasonable estimate of the fraction of adversary attacks that are made. Thus, the frequency of adversary attacks can be estimated from the number of adversary attacks that are detected.

Function 5. Plan a range of responses.

Any organization relying on information systems should have a number of routine information systems security activities (e.g., security features that are turned on, security procedures that are followed). But when attack is imminent (or in process), an organization could prudently adopt additional security measures that during times of non-attack might not be in effect because of their negative impact on operations. Tailoring in advance a range of information systems security actions to be taken under different threat conditions would help an organization plan its response to any given attack.

For example, a common response under attack is to drop non-essential functions from a system connected to the network so as to reduce the number of routes for penetration. A determination in advance of what functions count as non-essential and under what circumstances such a determination is valid would help facilitate an orderly transition to different threat conditions, and would be much better than an approach that calls for dropping all functionality and restoring only those functions that people using the system at the time complain about losing. Note that such determinations can be made only from an operational perspective rather than a technical one, a fact that points to the essential need for an operational architecture in the design of C4I systems.

The principle underlying response planning should be that of "graceful degradation"; that is, the system or network should lose functionality

---

[15]Furthermore, actions can be taken to minimize the possibility that adversaries might be able to obtain such information. For example, passing the information to the tested installation using non-electronic means would eliminate the possibility that an adversary monitoring electronic channels could obtain it.

gradually, as a function of the severity of the attack compared to its ability to defend against it.[16] This principle stands in contrast to a different principle that might call for the maintenance of all functionality until the attack simply overwhelms the defense and the system or network collapses. The latter principle is tempting because reductions in functionality necessitated for security reasons may interfere with operational ease of use, but its adoption risks catastrophic failure.

It is particularly important to note that designing a system for graceful degradation depends on system architects who take into account the needs of security (and more generally, the needs of coping with possible component failures) from the start. For example, the principle of graceful degradation would forbid a system whose continued operation depended entirely on a single component remaining functional, or on the absence of a security threat.

This principle is often violated in the development of prototypes. It is often said that "it is necessary for one to crawl before one can run," i.e., that it is acceptable to ignore security or reliability considerations when one is attempting to demonstrate the feasibility of a particular concept. This argument is superficially plausible, but in practice it does not hold water. It is reasonable for a prototype to focus only on concept feasibility, ignoring considerations of reliability or security, only if the prototype will be thrown away and a new architecture is designed and developed from scratch to implement the concept. Budget and schedule constraints usually prevent such new beginnings, and so in practice the prototype's architecture is never abandoned, and security or reliability considerations must be addressed in the face of an architecture that was never designed or intended to support them.

Function 6. Coordinate defensive activities throughout the enterprise.

Any large, distributed organization has many information systems and subnetworks that must be defended. The activities taken to defend each of these systems and networks must be coordinated because the distributed parts have interconnections and the security of the whole organization depends on the weakest link. Furthermore, it is important for dif-

---

[16]Of course, graceful degradation assumes an ability to detect an attack and make adjustments to system operation and configuration in near-real time. It is possible that in preparation for an attack, a clever opponent will be able to plant initially undetected "Trojan horses" that can be activated when the attack begins in earnest, or other programs that can operate covertly, making it hard for the defender to respond to an attack that is ongoing. This fact does not negate the utility of the design philosophy, but it does point out that graceful degradation cannot solve all security problems.

ferent parts of organizations to be able to learn from each other about vulnerabilities, threats, and effective countermeasures.

Function 7. Ensure the adequacy, availability, and functioning of public infrastructure used in systems (a step that will require cooperation with commercial providers and civilian authorities).

Few networks are built entirely using systems controlled by the organization that relies on them. Therefore organizations (including DOD) are required to work cooperatively with the owners of the infrastructure they rely on and relevant authorities to protect them.

Function 8. Include security requirements in any specification of system or network requirements that is used in the acquisition process.

Providing information systems security for a network or system that has not had security features built into it is enormously problematic. Retrofits of security features into systems not designed for security invariably leave security holes, and procedural fixes for inherent technical vulnerabilities only go so far.

Perhaps more importantly, security requirements must be given prominence from the beginning in any system conceptualization. The reason is that security considerations may affect the design of a system in quite fundamental ways, and a designer who decides on a design that works against security should at least be cognizant of the implications of such a choice. This function thus calls for information systems security expertise to be integrally represented on design teams, rather than added later.

Note that specification of the "Orange Book" security criteria[17] would be an insufficient response to this function. "Orange Book" criteria typically drive up development times significantly, and more importantly, are not inherently part of an initial requirements process and do not address the security of networked or distributed systems.

Function 9. Monitor, assess, and understand offensive and defensive information technologies.

Good information systems security requires an understanding of the types of threats and defenses that might be relevant. Thus, those responsible for information systems security need a vigorous ongoing program

---

[17]*Department of Defense Trusted Computer System Evaluation Criteria* (the Orange Book). December 1985. DOD 5200.28-std; supersedes CSC-STD-001-83, dated August 15, 1983.

to monitor, assess, and understand offensive and defensive information technologies. Such a program would address the technical details of these technologies, their capability to threaten or protect friendly systems, and their availability.

Function 10. Advance the state of the art in defensive information technology (and processes) with research.

Although much can usually be done to improve information systems security simply through the use of known and available technologies, "bug fixes," and procedures, better tools to support the information systems security mission are always needed. In general, such improvements fall into two classes (which may overlap). One class consists of improvements so that tools can deal more effectively with a broader threat spectrum. A second class, equally important, provides tools that provide better automation and thus can solve problems at lower costs (costs that include direct outlays for personnel and equipment and operational burdens resulting from the hassle imposed by providing security).

Similar considerations apply to processes for security as well. It is reasonable to conduct organizational research into better processes and organizations that provide more effective support against information attacks and/or reduce the impediments to using or implementing good security practices.

Function 11. Promote information systems security awareness.

Just as it is dangerous to rely on a defensive system or network architecture that is hard on the outside and soft on the inside, it is also dangerous if any member of an organization fails to take information systems security seriously. Because the carelessness of a single individual can seriously compromise the security of an entire organization, education and training for information systems security must be required for all members of the organization. Moreover, such education and training must be systematic, regarded as important by the organization (and demonstrated with proper support for such education and training), and undertaken on a regular basis (both to remind people of its importance and to update their knowledge in light of new developments in the area).

Function 12. Set security standards (both technical and procedural).

Security standards should articulate in well-defined and actionable terms what an organization expects to do in the area of security. They are therefore prescriptive. For example, a technical standard might be "all passwords must be eight or more characters long, contain both letters and numbers, be pronounceable, and not be contained in any dictionary," or "all electronic communications containing classified information must be encrypted with a certain algorithm and key length." A standard involving both technical and procedural measures might specify how to revoke cryptographic keys known to have been compromised. Furthermore, security standards should be expected to apply to all those within the organization. (For example, generals should not be allowed to exercise poorer information systems security discipline than do captains, as they might be tempted to do in order to make their use of C4I systems easier.)

Function 13. Develop and use criteria for assessing security status.

Information security is not a one-shot problem, but a continuing one. Threats, technology, and organizations are constantly changing in a spiral of measures and countermeasures. Organizations must have ways of measuring and evaluating whether they have effective defensive measures in place. Thus, once standards are put in place, the organization must periodically assess the extent to which members of the organization comply with those standards, and characterize the nature of the compliance that does exist.

Metrics for security could include number of attacks of different types, fraction of attacks detected, fraction of attacks repelled, damage incurred, and time needed to detect and respond to attacks. Note that making measurements on such parameters depends on understanding the attacks that do occur—because many attacks are not detected today, continual penetration testing is required to establish such a baseline.

One example of such monitoring is the efforts the National Security Agency (NSA) makes to ensure that cryptographic devices are being used. NSA can detect if any U.S. military communicators shut off cryptographic communications security (COMSEC) devices, and provides appropriate feedback to the relevant commands.

3.4 RESPONSIBILITY FOR INFORMATION SYSTEMS SECURITY IN DOD

The responsibility for information systems security within the Department of Defense is distributed through the entire organization, including both civilian and military components. The Assistant Secretary of Defense for Command, Control, Communications, and Intelligence (C3I) is the principal staff assistant to the Secretary of Defense for C3I and information management and information warfare matters and is the Chief Information Officer for the DOD. Other Office of the Secretary of Defense components with some connection to information systems security include the Defense Information Systems Agency (DISA), the Defense Advanced Research Projects Agency (DARPA), the National Security Agency (NSA), the Defense Intelligence Agency (DIA), and DOD's federally funded research and development centers, such as MITRE, the Institute for Defense Analyses, and RAND. Each of the military services and the combatant commands have one or more activities focusing on information systems security, as does the Joint Staff.

Organizations of particular relevance to the DOD-wide issues related to information systems security include the following:

- *The Defense-wide Information Assurance Program*, which was established in January 1998 to provide a "common framework and central oversight necessary to ensure the protection and reliability of the [Defense Information Infrastructure]."[18] The program's goal is to change the way DOD and its various agents look at information assurance, from a technical issue to an operational readiness issue. It will look at new tools (e.g., better systems) and techniques (e.g., vulnerability assessments, red team testing) to monitor and deter attacks on defense information systems.

- *The Defense Advanced Research Projects Agency (DARPA)*, which undertakes a large part of the DOD effort in basic R&D for information security. DARPA's efforts, located in its Information Technology Office (Information Survivability) and in the Information Systems Office (Information Assurance), are coordinated with NSA and DISA through a memorandum of understanding. The mission of the Information Assurance Program is to "develop security and survivability solutions for the Next Generation Information Infrastructure that will reduce vulnerability and allow increased interoperability and functionality."[19] The program's objectives include architecture and infrastructure issues, preventing, deterring, and responding to attacks, and managing security systems. Its goal is to "create the security foundation" for the Defense Information Infrastructure and future military C4I information systems.

- *The National Security Agency (NSA)*, which develops cryptographic and other information systems security techniques to protect sensitive (classified and unclassified) U.S. communications and computer systems associated with national security.[20] For many years, the NSA produced link encryptors that were used to protect data during communications.

---

[18]Remarks made by the Deputy Secretary of Defense, John J. Hamre, in his "Statement Before the Senate Armed Services Committee, Information Systems: Y2K & Frequency Spectrum Reallocation," June 4, 1998.

[19]See the Defense Advanced Research Projects Agency Information Assurance home page online at <http://web-ext2.darpa.mil/iso/ia/>.

[20]See the NSA's INFOSEC page online at <http://www.nsa.gov:8080/isso/>.

As the boundary between communications and computing has blurred, however, the NSA has broadened its mission to include information security rather than simply the more narrow communications security. Today, information protection activities are found within the Information Systems Security Organization, and this component of NSA houses considerable information security expertise.

- *The Defense Information Systems Agency (DISA)*, which serves as the manager for the Defense Information Infrastructure. In this role, DISA helps to "protect against, detect and react to threats to" the Defense Information Infrastructure and DOD information sources.[21] The INFOSEC Program Management Office coordinates all information security activities for DISA by providing technical and product support as well as INFOSEC education throughout the DOD. In addition, DISA's chief information officers' Information Assurance Division focuses on the implementation of information assurance by developing effective security policy and processes and establishing training and awareness program.[22] DISA also hosts the Joint Task Force on Computer Network Defense (Box 3.4), which is intended to work in conjunction with the unified military commands, the military services, and other Department of Defense agencies to defend DOD networks and systems against intrusions and other attacks.

- *The Joint Command and Control Warfare Center*, which is charged with providing direct tactical and technical analytical support for command and control warfare to operational commanders. The Joint Command and Control Warfare Center supports the integration of operations security, psychological operations, military deception, and electronic warfare and destruction throughout the planning and execution phases of operations. Direct support is provided to unified commands, joint task forces, functional and service components, and subordinate combat commanders. Support is also provided to the Office of the Secretary of Defense, the Joint Staff, the services, and other government agencies. The Joint Command and Control Warfare Center maintains specialized expertise in command and control warfare systems engineering, operational applications, capabilities, and vulnerabilities.

3.5 THE INFORMATION SYSTEMS SECURITY THREAT

Reliable estimates of national-level threats to DOD C4I systems are hard to obtain, even in the classified literature. Unlike more traditional

---

[21] For further information, see the Defense Information Systems Agency home page online at <http://www.disa.mil>.

[22] For additional information, see the Defense Information Systems Agency INFOSEC Program Management Office home page online at <http://www.disa.mil/infosec/index.html>.

BOX 3.4 The Joint Task Force on Computer Network Defense

The mission of the Joint Task Force on Computer Network Defense (JTF-CND) is to coordinate and direct the defense of DOD computer networks and systems. Thus, it serves as the focal point within the Department of Defense for organizing a united effort to defend DOD computer networks and systems. The JTF-CND mission includes the coordination of DOD defensive actions with non-DOD government agencies and appropriate private organizations. The JTF-CND directly supports critical infrastructure protection as discussed in Presidential Decision Directive 63 and in the Joint Vision 2010 notion of full spectrum dominance. The specific functions to be provided by the JTF-CND are as follows:

- Determine when system(s) are under strategic computer network attack, assess the impact on military operations and capabilities, and notify National Command Authorities and the user community.
- Coordinate and direct appropriate DOD actions to stop computer network attack, contain damage, restore functionality, and provide feedback to user community.
- Develop contingency plans, tactics, techniques, and procedures to defend DOD computer networks; support deliberate planning in the unified and specified commands for same.
- Assess the effectiveness of defensive actions, and maintain a current assessment of operational impact on DOD.
- Coordinate as required with national communications systems, the National Infrastructure Protection Center, DOD law enforcement agencies, DOD counterintelligence organizations, civilian law enforcement, other interagency partners, the private sector, and allies.
- Monitor the status of DOD computer networks.
- Monitor Computer Emergency Response Team alerts, warnings, and advisories, and provide input to and monitor indications and warnings (I&W) reporting.
- Participate in joint training exercises to conduct computer network defense.
- Coordinate with Defense-wide Information Assurance Program (DIAP) and Critical Asset Assurance Program (CAAP) authorities to ensure JTF-CND compliance with wider information assurance policy and initiatives.
- Provide the intelligence community with priority intelligence requirements for collection and I&W requirements for potential attacks against DOD computers and networks.
- Subject to authority, direction, and control of the Secretary of Defense, provide information to and receive direction from the Chairman of

continues

> the Joint Chiefs of Staff, and provide liaison as required to the staff of the Office of the Secretary of Defense and the Joint Chiefs of Staff.
>
> At present, the commander of the JTF-CND is also the vice director of the Defense Information Systems Agency. The JTF-CND is co-located with and hosted by DISA in order to take advantage of the existing operational computer network capabilities of DISA's Global Operations and Security Center, the military services, and DOD agencies. Initial operational capability was scheduled for December 31, 1998. Full operational capability will be achieved when the JTF-CND is able to accomplish all baseline functions around the clock; DOD plans to achieve full operational capability approximately 180 days following initial operational capability.
>
> SOURCES: MITRE Corporation and DOD News Release No. 658-98, "Joint Task Force on Computer Network Defense Now Operational," December 30, 1998.

threats (where vehicles and weapons platforms could be counted and exercises observed), the information security threat requires comparatively little capital and resources that are easily concealed, as well as expertise with both civilian and military applications, and so it is difficult to estimate. Thus, threat estimates in this domain are necessarily more dependent on human judgment, with all of the subjectivity and uncertainty thereby implied.

Essentially all nations with hostile intent toward the United States have the financial resources and the technological capability to threaten U.S. C4I systems. Because the costs of equipment to threaten U.S. C4I systems are small and the knowledge is available worldwide, non-state groups (e.g., terrorist groups or domestic hackers) can also pose a threat.

For these reasons, prudent planning dictates a serious DOD response to such potential threats, even if they have not yet been part of a concerted national attack on the United States.

3.6 TECHNICAL ASSESSMENT OF C4I SYSTEM SECURITY

The available evidence from exercises that the committee observed (e.g., Blue Flag 98-2) or received briefings on (e.g., Eligible Receiver) show that security at all levels, from the national down to the platform-level command, in today's fielded systems is insufficient. The security in today's fielded military systems is weak, and weaker than it need be, as illustrated by the following examples of behavior and practices that the committee observed or heard:

- Individual nodes are running commercial software with many known security problems. Operators use little in the way of tools for finding these problems, to say nothing of fixing them.
- Computers attached to sensitive command and control systems are also used by personnel to surf Web sites worldwide, raising the possibility that rogue applets and the like could be introduced into the system.[23]
- Units are being blinded by denial-of-service attacks, made possible because individual nodes were running commercial software with many known security problems.
- IP addresses and other important data about C2 systems can be found on POST-IT notes attached to computers in unsecured areas, making denial of service and other attacks much easier.
- Some of the networks used by DOD to carry classified information are protected by a perimeter defense. As a result, they exhibit all of the vulnerabilities that characterize networks protected by perimeter defenses.[24]

3.7 FINDINGS

Finding S-1: Protection of DOD's information and information systems is a pressing national security issue.

DOD is in an increasingly compromised position. The rate at which information systems are being relied on outstrips the rate at which they are being protected. Also, the time needed to develop and deploy effective defenses in cyberspace is much longer than the time required to develop and mount an attack. The result is vulnerability: a gap between exposure and defense on the one hand and attack on the other. This gap is growing wider over time, and it leaves DOD a likely target for disruption or pin-down via information attack.

Finding S-2: The DOD response to the information systems security challenge has been inadequate.

As noted in section 3.6, the committee observed in its field visits a variety of inadequate responses to the security problem. Within the DOD, the National Security Agency is the primary repository of expertise with

---

[23]An "applet" is an application supplied by a host Web site that can be run locally. Thus, connecting to a Web site supplying a rogue applet can result in the running of a hostile program on the system viewing that Web page.

[24]It is ironic that the use of a perimeter defense for a C4I network is inconsistent with the more stringent rules for protecting classified data in physical environments. For example, the storage of classified documents requires a safe in a room that is fitted with an alarm.

respect to information systems security, and this repository is arguably the largest and best in the world. Nevertheless, DOD has been unable to translate this expertise into adequate information assurance defenses except in limited areas (primarily the supply of cryptographic devices). For example, the committee observed in one exercise NSA personnel working in intelligence roles and in support of an information warfare attack cell. The information warfare defensive cell, however, did not use NSA-supplied tools and was not directly supported by NSA personnel.

Many field commanders told the committee that "cyberspace is part of the battlespace," and several organizations within the DOD assert that they are training "C2/cyber warriors." But good intentions have not been matched by serious attention to cyberspace protection. Soldiers in the field do not take the protection of their C4I systems nearly as seriously as they do other aspects of defense. For example, information attack red teams were a part of some exercises observed by the committee, but their efforts were usually highly constrained for fear that unconstrained efforts would bring the exercise to a complete halt. While all red teams operate under certain rules of engagement established by the "white teams" that oversee exercises, the information attack red teams appeared to the committee to be much more constrained than was appropriate. In one exercise, personnel in an operations center laughed and mistakenly took as a joke a graphic demonstration by the red team that their operations center systems had been penetrated.

One particularly problematic aspect of the DOD response to information systems security is its reliance on passive defense. As noted above, passive defense does not impose a meaningful penalty against an opponent, and thus the opponent is free to probe until he or she finds a weak spot in the defense. This reliance on passive defense is not a criticism of DOD; rather, it is an unavoidable consequence of a high-level policy decision made by the U.S. government that retaliation against cyber-attackers is not to be controlled or initiated by DOD; nevertheless, the committee is compelled to point out that this policy decision has a distinctly negative consequence for the security of DOD C4I systems.

On the technology side, the development of appropriate information systems security tools has suffered from a mind-set that fails to recognize that C4I systems are today heavily dependent on commercial components that often do not provide high levels of security. It may be true that the most secure systems are those that are built from scratch with attention from the start paid to security; in essence, this is the philosophy on which DOD's *Trusted Computer System Evaluation Criteria* are based.[25] But in prac-

---

[25]*Department of Defense Trusted Computer System Evaluation Criteria* (the Orange Book). December 1985. DOD 5200.28-std; supersedes CSC-STD-001-83, dated August 15, 1983.

tice, system builders must obtain security from whatever is provided by COTS products, security that is admittedly inadequate against the best efforts of world-class adversaries but that would improve security against less sophisticated threats. Because the National Security Agency has focused its efforts to date on the "build from scratch" philosophy, real-world military C4I systems built on commercial components have very little effective security and low assurance that they will work under real-world attacks by sophisticated opponents.

DOD efforts in information systems security have also focused a great deal of attention on high-assurance multilevel security. Multilevel security mechanisms seek to prevent a hostile piece of software from leaking high-level (e.g., secret) information to low-level (e.g., uncleared) users. While hostile "Trojan horse" software is certainly a real and important threat, it is far from the most serious problem facing command and control systems today. For example, denial-of-service attacks represent a serious threat, not least because such attacks may be the easiest to conduct. Moreover, the U.S. computer industry has not found sufficient demand, either from the DOD or elsewhere, for multilevel security-qualified systems.[26] Multilevel security may still be needed for certain specialized C4I applications, but from the standpoint of meeting the broad demands for security it has not proven to be a commercially viable approach.

By contrast, the commercial sector has taken a largely pragmatic approach to the problem of information systems security. The C4I security practices that the committee observed in many of its site visits were far inferior to the standard set by the best commercial practices for information systems security (e.g., those found in the banking industry) or the best practices in DOD. Given the importance of DOD C4I systems to the national security and the sensitivity of the information handled in those systems, the committee would have expected DOD C4I security practices, in general, to reach a higher standard than was found. Also, the committee observed a number of instances in which the adoption of existing good

---

[26]At one time, the U.S. computer industry was preparing at its own expense high-assurance multilevel security systems for use by DOD. These systems failed to make the transition from development project to commercial product. Perhaps the best example of such a system is Digital Equipment Corporation's VAX Virtual Machine Monitor security kernel. This project was canceled, apparently for commercial reasons, in 1991. (See Paul A. Karger, Mary Ellen Zurko, Douglas W. Bonin, Andrew H. Mason, Clifford E. Kahn. 1991. "A Retrospective on the VAX VMM Security Kernel," *IEEE Transactions on Software Engineering* 17(11): 1147-1165.) The committee is aware of no similar systems on the horizon today. One major reason for the lack of demand for such systems is that the time to market of multilevel security-qualified systems is so long that the functional capabilities of these systems have been superseded many times over by other non-multilevel security systems by the time they are available.

technologies and practices would greatly improve information systems security. Because these best practices have not been adopted for military use, the protection of C4I cyberspace is worse than it need be, and there is a large gap between the security that is reasonably possible today and the security that is actually in place.

An analogy that illustrates a more pragmatic approach is to view the threat as a pyramid. A large percentage of the low-level threats at the base of the pyramid can be handled with readily available tools. This keeps the "ankle biters" out. The apex of the pyramid represents that small percentage of "professionals" with largely unlimited resources that, given time, will be able to penetrate any defense. The middle levels, then, are the ones that benefit most from concentrated system design work.

3.8 RECOMMENDATIONS

The committee believes that information systems security—especially in its operational dimensions—has received far less attention and focus than the subject deserves in light of a growing U.S. military dependence on information dominance as a pillar of its warfighting capabilities.

At the highest level of abstraction, the committee believes that *DOD must greatly improve the execution of its information systems security responsibilities.* The same military diligence and wisdom that the U.S. military uses to defend physical space can and must be applied to defend the cyberspace in which C4I systems operate. For example, the principle of defense in depth is a time-honored one, whose violation has often led to military disaster (e.g., the Maginot line).

This is easier said than done. The defense of physical spaces and facilities has a long history, while cyberspace is a new area of military operations. In cyberspace, boundaries are fluid, control is distributed and diffuse, and most of what occurs is invisible to the defender's five senses without appropriate augmentation. As a result, analogies between physical space and cyberspace cannot be perfect, and may even be misleading. Nevertheless, a goal of achieving "cyber-security" for C4I systems comparable to what can be achieved with physical security for physical facilities and spaces is a reasonable one that the DOD should strive to meet.

One critical aspect of improving information systems security is changing the DOD culture, especially within the uniformed military, to promote an information systems security culture. Organizational policies and practices are at least as important as technical mechanisms in providing information systems security. Policies specify the formal structures, ensure responsibility and accountability, establish procedures for deploying and using technical means of protection and assigning access privileges, create sanctions for breaches of security at any level of the organization, and require training in the privacy and security practices of an organization. Thus, the organizational issues relating to how to ensure the appropriate use of information systems security technologies are critical.

The culture of any organization establishes the degree to which members of that organization take their security responsibilities seriously. With a culture that values the taking of the offensive in military operations, the military may well have difficulty in realizing that defense against information attack is a more critical function than being able to conduct similar operations against an adversary, and indeed is more difficult and requires greater skill and experience than offensive information operations.

For example, the committee observed the 609th Information Warfare Squadron in action during the Blue Flag 98 exercise. The 609th Squadron had split responsibilities: it was responsible for both red team (attacking) and blue team (defending) information activities. The defensive cell performed its duties admirably, yet was overwhelmed by its red team counterpart. (For example, the red team was able to download the air tasking order before it was transmitted.) In asking about the composition of the two teams, committee members were told that blue team defensive duty and experience were a prerequisite for participation on the red team.[27]

The notion that less experienced personnel first perform the defensive function and more experienced ones perform the offensive function is counter to normal practice in other settings. For example, the National Security Agency requires code-breaking experience before an analyst can begin to develop encryption algorithms. In general, the rule of good practice in information systems security is that the most experienced people serve the vital protection function.

In all instances of which the committee is aware, large organizations that take information systems security seriously have leadership that emphasizes its importance. Top-level commitment is not sufficient for good security practices to be put into place, but without it, organizations will drift to do other things that appear more directly related to their core missions. Thus, senior DOD leadership must take the lead to promote information systems security as an important cultural value for DOD.

---

[27]It can be argued that it is desirable to train against the most experienced adversaries. Indeed, experience at the National Training Center in which units in training are routinely overwhelmed by an experienced and superbly trained opposing force is based on this point. But for operational purposes, the commander must decide where to deploy his best personnel—and the committee believes cyber-defense warrants the very best. Because units fight as they train, the committee believes that the most experienced personnel should be involved as defenders in exercises, too. (An additional point is that the red-team threat so far overmatched the defense that red-team sophistication was never required.)

Senior leadership is responsible for two tasks: the articulation of policy for the department as a whole, and oversight to ensure that policy is being properly implemented.

In this regard, the committee was encouraged by conversations with senior defense officials, both civilian and military, who appear to take information systems security quite seriously. Nevertheless, these officials have a limited tenure, and the issue of high-level attention is a continuing one.

A second obstacle to the promulgation of an information systems security culture is that good security from an operational perspective often conflicts with doing and getting things done. And because good information systems security results in nothing (bad) happening, it is easy to see how the can-do culture of DOD might tend to devalue it.

Finally, it is important to note that DOD must protect both classified and unclassified information. While DOD has a clear legislative mandate to protect both types of information, DOD treats the protection of classified information much more seriously than the protection of unclassified information.

The first step is to take action now. Exercises such as Eligible Receiver have served as a "wake-up" call for many senior DOD leaders, both civilian and military. The perception at the highest levels of leadership that the information systems security problem is big, urgent, and real must translate quickly into actions that can be observed in the field.

One way of characterizing the committee's recommendations is that the DOD should adopt as quickly as is possible best commercial practices, which are in general far in advance of what the committee observed with fielded C4I systems. It is essential that security requirements be considered from the very beginning of each program and not postponed until later, which inevitably causes either major cost increases or the requirements to be diluted or eliminated. As a next goal DOD must then attempt to advance the state of the art in each of these areas.

Finally, in an organization as large as DOD, recommendations must refer to concrete actions and to specific action offices responsible for their execution. On the other hand, given an ongoing restructuring and streamlining within DOD, especially within the Office of the Secretary of Defense and the Joint Chiefs of Staff, the committee is reluctant to specify action offices with too much confidence or precision. Thus, its recommendations are cast in terms of *what* the committee believes should be done, rather than specifying an action office. The argumentation for each recommendation contains, where appropriate, a paragraph regarding a *possible* action office or offices for that recommendation, representing the committee's best judgment in that area. However, this action office (or offices) should be regarded as provisional, and DOD may well decide that a different action office is more appropriate given its organizational structure.

Recommendation S-1: The Secretary of Defense, through the Assistant Secretary of Defense for C3I and the Chairman of the Joint Chiefs of Staff, should designate an organization responsible for providing direct operational support for cyber-defense to commanders.

As noted above, defensive information operations require specialized expertise that may take years to develop. Thus, it is in the short run unrealistic to expect operational units to develop their own organic capabilities in this area. Because the committee believes that all operators and commanders during exercises and operations must be supported in the C4I defensive role by specialized experts serving in operations centers, it makes sense to organize units that can be deployed with forces that are dedicated to providing operational support. Providing such support also reinforces the commitment of DOD to this mission.

In its site visits, the committee observed limited resources devoted to providing operational support for the information systems security mission in some instances, such as the 609th Information Warfare Squadron at Blue Flag 98. But even in these instances (and they were not frequent), the defensive resources and efforts have been paltry compared to the magnitude and severity of the threat. The National Security Agency provides invaluable technical support, but for the most part does not appear to provide direct operational support to deployed units (or those on exercise). The services are beginning to pay more attention to the requirements of information systems security, and each has established an information warfare component, another promising development. But until the operators are brought into the picture in a central and visible manner, the security of fielded systems will remain inadequate.[28]

Only the Secretary of Defense has the necessary defense-wide purview of authority to designate and properly fund an appropriate organization to perform this function. The committee is silent on the appropriate executing organization, but notes that today the Joint Command and Control Warfare Center does do some of the things that the committee believes should be done in providing direct defensive support to commanders, although not on the scale that the committee believes is necessary. Furthermore, the Joint Task Force on Computer Network Defense is

---

[28]Today, NSA does provide significant signal intelligence support to field commanders. Whether or not it is the NSA that is tasked with providing defensive support to operational commanders, this NSA role with respect to signal intelligence suggests the feasibility of such a role for some organization.

charged with operational defensive responsibilities; it remains to be seen whether this organization provides adequate defensive support to commanders in the field.

Recommendation S-2: The Secretary of Defense should ensure that adequate information system security tools are available to all DOD civilian and military personnel, direct that all personnel be properly trained in the use of these tools, and then hold all personnel accountable for their information system security practices.

Accountability for upholding the values of an organization is an essential element of promulgating a culture. Once senior leaders have articulated a department-wide policy for information systems security and provided personnel with appropriate tools, training, and resources, it is necessary to develop well-defined structures with clear lines of responsibility.

Policies require procedures to translate their intent and goals into everyday practices, which may vary somewhat across departments. The most important aspect of such procedures is that authority and responsibility for implementation must be clearly assigned and audited by higher authority. In addition, units within the organization need procedures to determine the proper access privileges to an information system for individuals. Furthermore, privileges once determined must be established responsively (e.g., a new user needs certain privileges granted quickly in order to perform his or her job, and users who have been compromised must have their privileges revoked quickly).

In addition to the necessary policies and procedures, accountability within DOD rests on several pillars, including:

- *Education and training.* All users of information and C4I systems must receive some minimum level of training in relevant security practices *before* being granted access to these systems. Refresher courses are also necessary to remind long-time users about existing practices and to update them on changes to the threat. Note also that training activities for information systems security can be seen as a disruptive and unnecessary intrusion into the already busy schedule of personnel.
- *Incentives, rewards, and opportunities for professional advancement.* For security to be taken seriously, people within the organization must see the benefits and costs of compliance with good security practices. For example, promotions and an upward career path should be possible for specialists in information systems security, understanding that unless pay scales are changed, the lure of the private sector may prove irresistible for many individuals. Personnel who demonstrate extraordinary diligence

or performance under information attack should be eligible for special recognition (e.g., cash awards, medals).

- *Individual and unit-based measures of performance.* Military and civilian personnel should have an information security component as part of their performance ratings. Units should be rated with respect to their information security practices in exercises.
- *Sanctions.* The other side of rewards is sanctions. Sanctions for violations of good information systems security practice must be applied uniformly to all violators. Experience in other organizations indicates that if security practices are violated and no response follows, or if sanctions are applied, but only irregularly, after a long delay, or with little impact on perpetrators, the policy regime promoting security is severely undermined, and its legitimacy is suspect. Commanders and high-ranking officials, in particular, are often willing to compromise security practices for their own convenience and ease of use, and may not give the subject due attention in their oversight roles. It is thus not unreasonable that system administrators and their commanders, given the necessary tools, training, and resources, be held accountable for keeping systems configured securely and maintaining good operational security practices with respect to information systems security.[29]

Because this recommendation calls for an across-the-board cultural change within DOD, many different offices must be involved. The senior leadership within the department—the Secretary of Defense—must take responsibility for a department-wide policy on information systems security. The service secretaries and their military chiefs of staff must develop policies that tie performance on information systems security issues to appropriate sanctions and rewards. Given the National Security Agency's traditional role in providing tools for information security, the National Security Agency is probably the most appropriate agency to identify available tools that are practically usable by DOD personnel at all levels of seniority and irrespective of specialized expertise (i.e., they should be usable by tank commanders as well as C4I specialists). Military departments and the Office of the Secretary of Defense must take steps to instruct military and civilian personnel, respectively, in the use of these tools.

---

[29]For example, the Army has explored the possibility of security regulations that would make base commanders and systems operators liable for information systems intrusions under the military's Uniform Code of Military Justice. See Elana Varon, "Army to Hold Commanders and Sysops Liable for Hacks," *Federal Computer Week*, February 2, 1998.

Recommendation S-3: The Secretary of Defense, through the Assistant Secretary of Defense for C3I, the Chairman of the Joint Chiefs of Staff, and the CINCs, should support and fund a program to conduct frequent, unannounced penetration testing of deployed C4I systems.

As noted above, a continuing search for technical, operational, and procedural vulnerabilities in a network or system is essential, especially for those that are operating in an exercise or in an actual deployment. (An example of such a search is the communications security monitoring undertaken by the National Security Agency. In other domains such as base security, unscheduled red team visits are not uncommon.) Such tests should be conducted at a level consistent with a high-grade threat, and must be conducted against different C4I assets. These red team or tiger team probes would be unscheduled and conducted without the knowledge of the installation being probed; furthermore, the teams conducting would report to and be under the direction of parties that are separate from those of the installation being tested. Information gleaned from these probes should be passed to cognizant authorities within the DOD and the administrator of the network penetrated; if a penetration is successful where the implementation of a known fix would have stopped the penetration, the commander of the installation and the administrator should be sanctioned. Note the critical focus on C4I systems operating in a "full-up" mode, rather than on individual C4I components.

A second important element of penetration testing is for the installation itself—probably under the technical direction of the on-site system administrator—to conduct or request its own penetration testing. Information on successful penetrations conducted under these auspices should still be shared with cognizant DOD authorities, but in order to encourage installation commanders to conduct such testing on their own, sanctions should not be applied to vulnerabilities that are discovered.

In the area of DOD-wide penetration testing, the Assistant Secretary of Defense for C3I has the authority to direct such testing. The CINCs, especially the U.S. Atlantic Command as the force provider, have operational responsibilities, and the Joint Chiefs of Staff must cooperate in the promulgation of policy in this area because such testing has a direct impact on operational matters. The committee also notes that the Information Warfare Red Team of the Joint Command and Control Warfare Center in San Antonio, Texas,[30] was created to improve the readiness posture of the DOD by identifying vulnerabilities in information systems and vulnerabilities caused by use of these information systems and then demon-

---

[30]For additional information about the Information Warfare Red Team, see the OSD Web page online at <http://www.acq.osd.mil/at/a6-iwrt.htm>.

strating these vulnerabilities to operators and developers (sometimes as part of the opposition force in exercises). The Information Warfare Red Team was initiated in 1995 and is sponsored jointly by the Deputy Under Secretary of Defense for Acquisition and Technology, the Office of the Assistant Secretary of Defense for C3I, and the Joint Staff Operations Directorate. Establishing the Information Warfare Red Team is an important step in the right direction to support the intent of this recommendation, but the scale of the activities undertaken by the Information Warfare Red Team is incommensurate with the much larger need for such testing.

Recommendation S-4: The Assistant Secretary of Defense for C3I should mandate the immediate department-wide use of currently available network and configuration management tools and strong authentication mechanisms.

Many information vulnerabilities arise from improper system or network configuration.[31] For example, a given system on a network may have a modem improperly attached to it that is not known to the network administrator. It may be attached for the most benign of reasons, such as a programmer or an applications developer who needs off-hours access to the system to complete work on an application on time. But the very presence of such a device introduces a security hole through which penetrations may occur. Or, a firewall may be improperly configured to allow Web access for a certain system when in fact the system should only be able to transmit/receive e-mail. Default passwords and accounts may still be active on a given system, allowing adversaries inappropriate access. Foreign software may have been downloaded inadvertently for use on some system, software whose purpose is hostile.

A network/system administrator should know the configuration of the network/systems for which he is responsible. He or she should be able to find unauthorized modems, poor passwords, factory settings, and unpatched holes in operating systems. But because checking an operational configuration is very labor-intensive if done manually, configuration management and network assessment tools must be able to run under automated control on a continuous basis, alerting the administrator when variances from the known configuration are detected. Some tools are available to do configuration management and network assessment, as well as inspection tools that allow correct configurations to be in-

---

[31]As used here, system or network configuration does not refer to what is often called source code configuration management, but rather to administrator-determined settings for services to be made available to various users, and other such "run-time" configuration parameters.

spected. These tools are not perfect, but their widespread use would be a significant improvement over current DOD practice.

A second aspect of configuration control is more difficult to achieve. Good configuration control also requires that every piece of executable code on every machine carry a digital signature that is periodically checked as a part of configuration monitoring. Furthermore, code that cannot be signed (e.g., macros in a word processor) must be disabled until development indicates a way to sign it. Today, it is quite feasible to require the installation of virus-checking programs on all servers and to limit the ability of users to download or install their own software (though Java and Active-X applets do complicate matters to some extent). Census software or regular audits can be used to ensure compliance with such policies. However, no tool known to the committee and available today undertakes this task systematically.

Note that it is not practical to secure every system in the current inventory. It is probably unrealistic to develop and maintain tools that do thorough monitoring of the security configuration for more than two or three platforms (e.g., Windows NT and Sun UNIX). Over the long run, it may well be necessary to remove other systems from operational use, depending on the trade-offs between lower costs associated with maintaining fewer systems and greater security vulnerabilities arising from less diversity in the operating systems base.

Authentication of human users is a second area in which DOD practices do not match the best practices found in the private sector. Passwords—ubiquitously used within the DOD as an authentication device—have many well-known weaknesses. An adversary can guess passwords, or reuse a compromised password (e.g., one found in transit on a network by a "sniffer"), and can compromise a password without the knowledge of its legitimate user.

A hardware-based authentication mechanism suffers from these weaknesses to a much lesser extent.[32] Because the mechanism is based on a physical piece of hardware, it cannot be duplicated freely (whereas passwords are duplicated when one person tells another a password). The hardware can be designed to be tamper-resistant, which increases the difficulty of duplicating it. Furthermore, because persistent (i.e., long-last-

---

[32]The device (e.g., a personal computer card) is enabled by a short password, usually called a PIN, entered by the user directly into the device. The device then engages in a secure and unforgeable cryptographic protocol with the system demanding the authentication; this protocol is much stronger than any password could be. The use of passwords is strictly local to the device and does not suffer from the well-known problems of passwords on networks, for example "sniffing" and playback attacks. This authentication depends on what you have (the device) together with that you know (the PIN).

ing) identifying information is never transmitted outside the piece of hardware, attacks to which password authentication is vulnerable (e.g., sniffing and playback attack) are essentially impossible. Hardware-based authentication is a highly effective method for authenticating communications originating from individuals. It also has particular value in the protection of remote access points (Box 3.5).

Biometric identifiers complement hardware-based authentication devices. Because biometric information is closely tied to the user, biometric identifiers serve a function similar to that of the personal identification number (PIN) that is used to activate the device. Biometric identifiers are based on some distinctive physical characteristics of an individual (e.g., a fingerprint, a voiceprint, a retinal scan); biometric authentication works by comparing a real-time reading of some biometric signature to a previously stored signature. Biometric authentication is a newer technology than that of hardware-based authentication; as such it is less well developed (e.g., slower, less accurate) and more expensive even as it promises to improve security beyond that afforded by PINs.

BOX 3.5 Protection of Remote Access Points

Remote access points pose particular vulnerabilities. A hostile user attempting to gain access to a computer on the premises of a U.S. command post, for example, must first gain physical entry to the facility. He also runs the risk of being challenged face to face in his use of the system. Thus, it makes far more sense for an adversary to seek access remotely, where the risk of physical challenge is essentially zero.

Strong authentication—whether hardware-based or biometric—is thus particularly important for protecting remote access points that might be used by individuals with home or portable computers. Some organizations (not necessarily within the DOD) protect their remote access points by using dial-back procedures[1] or by embedding the remote access telephone number in the software employed by remote users to establish a connection. Neither approach is adequate for protecting remote access points (e.g., dial-back security is significantly weakened in the face of a threat that is capable of penetrating a telephone switch, such as a competent military information warfare group), and their use does not substitute for strong authentication techniques.

---

[1]In a dial-back procedure, a remote user dials a specified telephone number to access the system. The system then hangs up and checks the caller's number against a directory of approved remote access telephone numbers. If the number matches an approved number, the system dials the user back and restores the connection.

Hardware-based authentication can also be used to authenticate all computer-to-computer communications (e.g., those using security protocols such as Secure Sockets Layer or IPSec). In this way, all communications carried in the network can be authenticated, not just those from outside a security perimeter. "Mutual suspicion" requiring mutual authentication among peers is an important security measure in any network.

The potential value of strong authentication mechanisms is more fully exploited when the authentication is combined with mechanisms such as IPSec or TCP wrappers that protect the host machines against suspicious external connections[33] and a fine-grained authorization for resource usage. For example, a given user may be allowed to read and write to some databases, but only to read others. Access privileges may be limited in time as well (e.g., a person brought in on a temporary basis to work on a particular issue may have privileges revoked when he or she stops working on that issue). In other words, the network administrator should be able to establish groups of users that are authorized to participate in particular missions and the network configured to allow only such interactions as necessary to accomplish those missions. Similarly, the network administrator should be able to place restrictions on the kinds of machine-to-machine interactions allowable on the network. This requires that the administrator have tools for the establishment of groups of machines allowed to interact in certain ways.

Some network management/configuration systems allow configuration control that would support fine-grained access controls. But most do not make it easy for a network administrator to quickly establish and revoke these controls.

Finally, the trend of today toward "single login" presents a dangerous vulnerability.[34] When a perimeter defense is breached, an adversary

---

[33]TCP wrappers protect individual server machines, whereas firewalls protect entire networks and groups of machines. Wrappers are programs that intercept communications from a client to a server and perform a function on the service request before passing it on to the service program. Such functions can include security checking. For example, an organization may install a wrapper around the patient record server physicians use to access patient information from home. The wrapper could be configured to check connecting Internet addresses against a predefined approved list and to record the date and time of the connection for later auditing. Use of wrapper programs in place of firewalls means that all accessible server machines must be configured with wrapper(s) in front of network services, and they must be properly maintained, monitored, and managed. See Wietse Venema. 1992. "TCP WRAPPER: Network Monitoring, Access Control and Booby Traps," pp. 85-92 in *Proceedings of the Third Usenix UNIX Security Symposium,* Baltimore, Md., September.

[34]"Single login" refers to the need of a user to log in (and authenticate himself) only once per session, regardless of how many systems he accesses during that session.

can roam the entire network without ever being challenged again to authenticate himself. A more secure arrangement would be for the network to support remote interrogation of the hardware authentication device by every system the user attempts to access, even though the user need only enter the PIN once to activate the device. In this way, every request to a computer, no matter where it is located on the network, is properly supported by strong evidence of the machine and the individual that is responsible for the request, allowing this evidence to be checked against the rules that determine who is allowed access to what resources.

Implementing this recommendation is not easy, but is well within the state of the art. A reader for a hardware authentication device in every keyboard and in every laptop (via personal computer-card slots) is very practical today.[35] In principle, even smart "dog tags" could be used as the platform for a hardware authentication device. However, the most difficult issue is likely to be the establishment of the public-key infrastructure for DOD upon which these authentication devices will depend. Biometric authentication devices are not practical for universal deployment (e.g., for soldiers in the field), but they may be useful in more office-like environments (e.g., command centers).

Since DOD increasingly relies on commercial technology for the components of C4I systems, engagement of commercial support for authentication is important to making this affordable. It should be possible to enlist strong industry support for a program to make better authentication more afforable if the program is properly conceived and marketed. Many commercial customers have very similar requirements, which are poorly met by existing security products. Thus, from a practical standpoint, the DOD's needs with respect to authentication are very similar to commercial needs.

Because this recommendation calls for DOD-wide action with respect to C4I systems, the Assistant Secretary of Defense for C3I must promulgate appropriate policy for the department. The information security policy is within the purview of the DOD's Chief Information Officer, who today is also the Assistant Secretary of Defense for C3I. Finally, given its history of involvement with information systems security, the National Security Agency is probably the appropriate body to identify the best available authentication mechanisms and configuration tools.

---

[35]The Fortezza card was an attempt by the DOD in the mid-1990s to promote hardware-based authentication. While the Fortezza program itself has not enjoyed the success that was once hoped for it, the fact remains that one of the capabilities that Fortezza provides—widespread use of hardware-based authentication—is likely to prove a valuable security tool.

Recommendation S-5: The Under Secretary of Defense for Acquisition and Technology and the Assistant Secretary of Defense for C3I should direct the appropriate defense agencies to develop new tools for information security.

Aligning DOD information security practice with the best practices found in industry today would be a major step forward in the DOD information security posture, but it will not be sufficient. Given the stakes of national security, DOD should feel an obligation to go further still. Going further will require research and development in many areas.

For example, tools for systematic code verification to be used in configuration monitoring are an area in which DOD-sponsored research and development could have high payoff in both the military and civilian worlds, as organizations in both worlds face the same problem of hostile code.

A second example involves fine-grained authorization for resource usage. Some network management/configuration systems allow configuration control that would support fine-grained access controls. But most do not make it easy for a network administrator to quickly establish and revoke these controls, and DOD-sponsored research and development in this area could have high payoff as well.

A third area for research and development is tools that can be used in an adaptive defense of C4I systems. Adaptive defenses change the configuration of the defense in response to particular types of attack. In much the same way that an automatic teller machine eats an ATM card if the wrong PIN is entered more than three times, an "adaptive" defense that detects an attack being undertaken through a given channel can deny access to that channel for the attacker, thus forcing him to expend the time and resources to find a different channel.

More sophisticated forms of adaptive defense might call for "luring" the attacker into a safe area of the system and manipulating his cyber-environment to waste his time and to feed him misleading information. For example, certain known security holes can be left unfixed, so that an attacker can have relatively easy access to the system through those holes. However, in fact, the information and system resources accessible through those holes are structured in such a way that they look authentic while providing nothing useful to the attacker. Deceptive defenses can force the attacker to waste time so that the defense has a greater opportunity to monitor the attacker and/or track the attacker's location and to take appropriate action. On the other hand, its long-term success presumes that the attacker cannot distinguish the holes left open deliberately from the ones unintentionally left open and that the defenders have the discipline to monitor the former; thus, such "deceptive" techniques cannot be regarded as anything more than a component of effective cyber-defenses.

A fourth area for research and development is biometrics. The basic technology and underlying premises of biometrics have been validated, but biometric authentication mechanisms are still sometimes too slow and too inaccurate for convenient use. (For example, they often take longer to operate than typing a password, and they sometimes result in false negatives (i.e., they reject a valid user fingerprint or retinal scan).) Broad user acceptance will depend both on more convenient-to-use mechanisms and on the integration of biometrics into the man-machine interface, such as a fingerprint reader in a mouse or keyboard.

Finally, research and development on active defenses is needed. Active defenses make attackers pay a price for attacking (whether or not they are successful), thus dissuading a potential attacker and offering deterrence to attack in the first place (an idea that raises policy issues as important as those associated with Recommendation S-7 (below). Passive information systems security is extremely important but against a determined opponent with the time and resources to conduct an unlimited number of penetration attempts against a passive non-responding target, the attacker will inevitably succeed. This area for research and development raises important policy issues that are discussed below. But the fact remains that even if policy allowed the possibility of retaliation, the tools to support such retaliation are wholly inadequate. Instruments to support a policy-authorized retaliation are needed in two areas:

- *Identification of an attacker.* Before any retaliatory action can be undertaken, the attacker must be identified in a reasonable time scale with a degree of confidence commensurate with the severity of that action. Today, the identification of an attacker is an enormously time-consuming task—even if the identification task is successful, it can take weeks to identify an attacker. And, it is often that considerable uncertainty remains about the actual identity of the attacker, who may be an individual using an institution's computer without the knowledge or permission of that institution. Note also that better tools for the accurate and rapid location of cyber-attackers would greatly assist law enforcement authorities in apprehending and prosecuting them.

- *Striking back against an attacker.* Once an attacker is identified, tools are needed to attack him or her. Many of the techniques employed against friendly systems can be used against an attacker as well, but all of these techniques are directed against computer systems rather than individual perpetrators. Furthermore, using these techniques may well be quite cumbersome for friendly forces (just as they are for attackers). However, the most basic problem in striking back is that from a technical perspective, not enough is known about what retaliation and active defenses might be.

Other possible research and development areas include secure composition of secure systems and components to support ad hoc (e.g., coalition) activities; better ways to configure and manage security features; generation of useful security specifications from programs; more robust and secure architectures for networking (e.g., requiring trackable, certificated authentication on each packet, along with a network fabric that denies transit to unauthenticatable packets); and automatic determination of classification from content.

Many agencies within DOD can conduct research and development for better information security tools, but a high-level mandate for such activity would help increase the priority of work in this area for such agencies. The National Security Agency and the Defense Advanced Research Projects Agency are the most likely agencies to develop better tools for information systems security. As noted above, better tools developed for DOD use are also likely to have considerable application in the commercial sector, a fact that places a high premium on conducting research and development in this area in an unclassified manner. Note that *Trust in Cyberspace* also outlines a closely related research agenda.[36]

Recommendation S-6: The Chairman of the Joint Chiefs of Staff and the service Secretaries should direct that a significant portion of all tests and exercises involving DOD C4I systems be conducted under the assumption that they are connected to a compromised network.

Because both threat and technology evolve rapidly, perfect information systems security will never be achieved. Prudence thus requires C4I developers and operators to assume some non-zero probability that any system will be successfully attacked, that some DOD systems have been successfully attacked, and that some C4I systems are compromised at any given moment. (A "compromised" system or network is one that an adversary has penetrated or disrupted in some way, so that it is to some extent no longer capable of serving all of the functions that it could serve when it was not compromised.) This pessimistic assumption guards against the hubris of assumed perfection. However, despite this assumption, most of the C4I systems connected to the compromised components should be able to function effectively despite local security failures.

---

[36]Computer Science and Telecommunications Board, National Research Council. 1999. *Trust in Cyberspace*, National Academy Press, Washington, D.C.

C4I systems should be designed and developed so that their functions and connectivity are easy to reconfigure under different levels of information threat. Critical functions must be identified in advance for different levels of threat (at different "INFOCONS") so that responses can occur promptly in a planned and orderly fashion. Note also that the nature of a mission-critical function may vary during the course of a battle.

C4I systems should be tested and exercised routinely under the assumption that they are connected to a compromised network. The capability of U.S. forces against an adversary is strongly dependent on the training they receive, and so C4I tiger teams playing in exercises involving C4I (i.e., every exercise) should be able to operate in a largely unconstrained mode (i.e., subject to some but not many limits). The lack of constraint is intended to stress friendly forces in much the same way that very well trained opposition forces such as those at the Army's National Training Center, the Air Force's Air Warfare Center, and the Navy's Fighter Weapons School stress units that exercise there. However, because the activities of entirely unconstrained tiger teams may prevent the test or exercise from meeting other training goals, some limits are necessary. (The portion of the test or exercise subject to the assumption of a compromised network should also be expected to increase, and the limits on tiger team activities relaxed, as friendly forces develop more proficiency in coping with information threats.) With tiger teams operating in this mode, every battlefield C4I user could be made conscious that his information may have been manipulated and that at any instant it might be denied.

Note that assuming a compromised network does *not* necessarily mean that the network cannot be used—only that it must be used with caution. For example, the network can be continually monitored for indications of anomalous activity, even if the network is nominally regarded as "secure." Network configurations can be periodically altered to invalidate information that the enemy may have been able to collect about the network. These steps would be analogous to periodic changes in tactical call signs that are used to identify units, an operational security measure that is taken to frustrate (or at least to complicate the efforts of) enemy eavesdroppers.

Doctrine should account for the possibility that a tactical network has been compromised or penetrated as well. In addition to continually taking preventive measures even when the network is not known to have been compromised, commanders must have a range of useful responses when a compromise or penetration is detected. This premise differs from today's operational choices, which are either to stay connected to everything or to disconnect and have nothing, with added exhortations to "be careful" when intrusions are detected. Finally, units must know how they will function when the only C4I available to them is unsecured voice communications.

In short, it is useful for the U.S. military to be trained in how to use its C4I systems and networks even if they have been compromised, as well as for the possibility that they will be largely unavailable for use at all.

Because this recommendation affects all operational deployments and exercises, both service and joint, a number of offices must take action. The Chairman of the Joint Chiefs of Staff should promulgate a directive that calls for such a policy in all joint exercises and operational deployments. And, because many C4I systems are owned and operated and controlled by the services, the services—perhaps through their training and doctrine commands—should establish doctrinal precepts for commanders to follow in implementing this policy.

Recommendation S-7: The Secretary of Defense should take the lead in explaining the severe consequences for U.S. military capabilities that arise from a purely passive defense of its C4I infrastructure and in exploring policy options to respond to these challenges.

Because a purely passive defense will ultimately fail against a determined attacker who does not pay a price for unsuccessful attacks, a defensive posture that allows for the possibility of inflicting pain on the attacker would bolster the security of U.S. C4I systems.[37] Today, a cyber-attack on U.S. C4I systems is regarded primarily as a matter for law enforcement, which has the lead responsibility for apprehending and prosecuting the attacker. DOD personnel may provide technical assistance in locating and identifying the attacker, but normally DOD has no role beyond that.

If an attack is known with certainty to emanate from a foreign power (a very difficult determination to make, to be sure) and to be undertaken by that foreign power, the act can be regarded as a matter of national security. If so, then a right to self-defense provides legal justification for retaliation. If the National Command Authorities (i.e., the President and the Secretary of Defense, or their duly authorized deputies or successors) decides that retaliation is appropriate, the remaining questions are those of form (e.g., physical or cyber) and severity (how hard to hit back). Under such circumstances, DOD would obviously play a role. However, DOD is legally prohibited from taking action beyond identification of a cyber-attacker on its own initiative, even though the ability of the United

---

[37]DOD is not alone in having to deal with the difficulties of a purely passive defense. But given the importance to the national security, the inevitable consequences of passive defense have immense significance for DOD.

States to defend itself against external threats is compromised by attacks on its C4I infrastructure, a compromise whose severity will only grow as the U.S. military becomes more dependent on the leverage provided by C4I.

From a national security perspective, the geographical origin of the attack matters far less than the fact that it is military C4I assets that are being attacked. Thus, the military desirability of cyber-retaliation to protect the nation's ability to defend itself should be clear. But the notion of cyber-retaliation raises many legal and policy issues, including issues related to constitutional law, law enforcement, and civil liberties.

The legal issues are most significant in peacetime—if the United States were actively engaged in conflict, the restraints on DOD action might well be relaxed. But the boundary between peacetime and conflict is unclear, especially if overt military hostilities (i.e., force on force) have not yet broken out but an adversary is probing in preparation for an attack. It is this time that poses the most peril, because DOD is constrained—because it is "officially" peacetime and yet an adversary may be gaining valuable advantage through its probes.

As a first step, DOD should review the legal limits on its ability to defend itself and its C4I infrastructure against information attack.[38] After such a review, DOD should take the lead in advocating changes in national policy (including legislation, if necessary) that amend the current "rules of engagement" specifying the circumstances under which force is an appropriate response to a cyber-attack against its C4I infrastructure. These rules of engagement would explicitly specify the nature of the force that could be committed to retaliation (e.g., physical force, cyber-attack), the damage that such force should seek to inflict, the authorizations needed for various types of response, the degrees of certainty needed for various levels of attack, the issues that would need to be considered in any response (e.g., whether the benefits of exploiting the source of an attack outweigh the costs of allowing that attack to continue), and the oversight necessary to ensure that any retaliation falls within all the parameters specified in the relevant legal authorizations.

The committee is not advocating a change in national policy with respect to cyber-retaliation. Indeed, it was not constituted to address the larger questions of national policy, i.e., whether other national goals do or do not outweigh the narrower national security interest in protecting its military information infrastructure, and the committee is explicitly silent

---

[38]Press reports indicate that DOD authorities are "struggling to define new rules for deciding when to launch cyber attacks, who should authorize and conduct them and where they fit into an overall defense strategy." See Bradley Graham, "Authorities Struggle with Cyberwar Rules," *Washington Post,* July 8, 1998, page A1.

on the question of whether DOD should be given the authority (even if constrained and limited to specific types and circumstances) to allow it to retaliate against attackers of its C4I infrastructure. But it does believe that DOD should take the lead in explaining the severe consequences for its military capabilities that arise from a purely passive defense, that DOD should support changes in policy that might enable it, perhaps in concert with law enforcement agencies, to take a less passive stance, and that a national debate should begin about the pros and cons of passive versus active defense.

The public policy implications of this recommendation are profound enough that they call for involvement at the highest levels of the DOD—the active involvement of the Secretary of Defense is necessary to credibly describe the implications of passive defense for C4I systems in cyberspace.

To whom should DOD explain these matters? Apart from the interested public, the Congress plays a special role. The reason is that actual changes in national policy in this area that enable a less passive role for DOD will certainly require legislation. Such legislation would be highly controversial, have many stakeholders, and would be reasonable to consider (let alone adopt) only after a thorough national debate on the subject.

4

Process and Culture

4.1 MANAGING CHANGE

The catch phrase used to capture the technology-driven transformation in almost all aspects of warfighting in the years ahead—the so-called revolution in military affairs—points to the magnitude of the institutional challenges associated with this transformation. Revolutions do not occur smoothly, nor do they succeed without significant breakage on many fronts. Revolutionary change and transformation are even more difficult when the institutions are steeped in proud histories and imbued with strong cultures. And, in the absence of an immediate crisis facing them, institutions are particularly challenged to transform themselves.

Although the military situation is different in major ways from that of the industrial sector, some useful guidance is available in the form of generally effective principles that have been learned from the revolutions currently under way in banking, retailing, the distribution industry, and a number of other commercial sectors. Six keys to success derived from a study of successful transformations in the commercial sector are the following:

- A consistent and clear driving vision;
- A set of supporting processes, drawing broadly on those affected by change and often using specific institutions, to refine and communicate the vision, to quantify and test its reality, and to translate it into implementable pieces;
- A persistent and constant in-place leadership cadre, driving an ongoing sense of urgency;

- The willingness and drive to reengineer any process, doctrine, or organization and to take risks;
- The willingness to allocate the funding necessary for change and to reprioritize budget allocations; and
- A commitment to align the measurement system across the hierarchy and in accordance with the vision.

Each of these items is discussed below in the context of DOD implementation of these principles. Metrics for the management measurement system are addressed in section 4.6.

4.1.1 Clear Vision for the Future

As noted in Chapter 1, Joint Vision 2010 reflects the top-level vision in the DOD of what is possible through the exploitation of C4I technology, and the services have each translated this top-level vision into a service-specific vision.

Today, the culture of the DOD regarding C4I systems and capabilities is in a state of transition, with senior military leadership becoming more broadly aware of information technology as an evolutionary force in doctrine and operations. This evolution is characterized by changes in doctrine, growth in new descriptive terminology, and substantial leadership investment in awareness. In short, the committee believes that the DOD has performed reasonably well in articulating a vision for the future.

4.1.2 Supporting Processes

In the course of its work, the committee encountered a number of efforts aimed at refining and quantifying the vision of advanced C4I systems and at learning and capturing the creative energies of the services and numerous supporting industries; these efforts included some of the exercises and experiments of several services, and demonstrations such as the Joint Warrior Interoperability Demonstrations. The DOD leadership has also approved a number of recent concept studies and organizations that aim to better understand and, where possible, quantify the contributions to military effectiveness that can be realized from effectively exploiting information technology. So-called "battle labs," along with numerous simulations, experiments, tests, and exercises, have contributed to a body of significant knowledge regarding the utility of advanced C4I systems. The committee found a large number of overlay offices and processes aimed at achieving jointness and interoperability, indicating at least some significant organizational acknowledgment of these matters.

Nevertheless, as is often the case in the evolution of any large enterprise, DOD's doctrinal and technical visionaries are far ahead of DOD's institutional reality in terms of bringing information technology to bear on current and future military needs. For example, the mere presence of offices and processes does not mean that the organization as a whole places a high priority on jointness and interoperability, and there is a big difference between a laboratory or an experiment that is fully joint in spirit and execution and one that is focused primarily on a specific service but with token involvement of other services at the edges.[1] And, in the information security area, the committee did not observe a comparable organizational acknowledgment.

Many factors have been blamed for the less-than-full realization of the potential impact of C4I on military operations. These factors include the following:

- Equipment, by law, must be purchased in the individual services.
- Time and tradition have created distinctive cultures within the services.
- In each service, promotion depends heavily on combat command experience.
- Congressional mandates have forced the DOD's suppliers to operate at greater and greater arm's length from those they serve.
- Traditional weapons change slowly, and the military has become accustomed to procurements that take many years to complete.
- Information technology in computers, communication, and sensors is changing at an exceptionally rapid pace.
- The military market for many commercial information technology products is comparatively small.

Numerous efforts are under way to deal with some of those individual causes. Yet none seems sufficient despite the prevalent answer to almost any question: "Yes, we're taking care of that." Because the issues are so diverse, it is necessary to aggregate the resources needed to deal with them, and thus an organizational approach to promoting change—much like Motorola University or the General Electric Crotonville school—seems a more promising approach.[2] In this context, the organization pro-

---

[1] An illustration would be an experiment that focuses on new doctrinal concepts for one service but that involves the other services based on their existing doctrine. Thus, for example, the Army experiments of Force XXI did not appear to take into account possible developments in Air Force expeditionary force doctrine. This point is an observation, not a criticism.

[2] Note that an actual physical institution is not necessarily needed. Modern techniques for managing independent collaboration through the use of software and communication techniques are well developed in industry, and experience makes it clear that close collaborative work and action can take place among separated groups and individuals. See J. Quinn, J. Baruch, and K. Zien. 1997. *Innovation Explosion,* The Free Press, New York, pp. 107-140.

vides an experimental context in which knowledge is captured at the point of action—because practitioners have the opportunity to codify their knowledge—and lessons learned become imperatives and lead to adaptation. If these lessons then link to doctrine, this learning process can be an institutional mechanism for responding to environmental change. It is critical to this approach that a strategic perspective be used, or else the knowledge management effort will degrade into one that creates only large repositories of reports. The role of communities of practice not only can affect the decision process (including key stakeholders and implementers) but also can begin to affect culture.

Consider, for example, General Electric's Crotonville Management School as a focus for then-CEO Jack Welch's continuing efforts to transform General Electric. Crotonville is far more than a typical management school. It is a key center of debate and refinement of the waves of strategic change that have made General Electric one of the world's most competitive and successful companies. It is both the source of refinement of the gospel and the place where it is debated and driven home across the General Electric management structure. The committee did not find any DOD entity analogous to the Crotonville school that might be the center of education and research aimed at driving the revolution in military affairs forward in a truly joint manner. To the best of the committee's knowledge no analogous transformation-driving institution exists within the military, particularly with respect to joint and/or combined operations.

4.1.3 Persistent Leadership Creating a Sense of Urgency

In the civilian world, chief executive officers and other key personnel can remain in place for as long as necessary to guide a substantial organizational change (i.e., time periods long enough to convince lower levels of the organization that waiting until the focus of management changes is not a viable option). Thus, CEOs and others can be chosen for their vision and commitment to change with the expectation that the individuals will endure. One military analog of such a leader was Admiral Hyman Rickover, an individual who personified a vision for the future with respect to the nuclear Navy and who could drive progress today, tomorrow, and every day beyond that.[3] A more recent example is Rear Admiral Wayne Meyer, who presided over the development and deployment of

---

[3]This is not to say that all of Admiral Rickover's decisions regarding the nuclear Navy were appropriate. For example, many have pointed to the fact that one of Admiral Rickover's legacies is the reliance on nuclear reactors whose design and performance fall far short of what would be possible with other nuclear reactor technologies. Nevertheless, it is undeniable that Rickover had a profound influence on keeping the Navy focused on developing nuclear propulsion technology.

AEGIS for almost 14 years.[4] It is also significant that AEGIS was developed more or less within all of the acquisition bureaucracy and constraints of the testing community.

But in the military, the long tenure of individuals is the exception—in general, short time limits on the tenure of driving visionaries within the DOD encourage the natural tendency of an organization to resist change. The DOD suffers from management turnover in the top military (and civilian) leadership that is much more frequent than turnover in the private-sector companies that have successfully effected major cultural changes. The committee met and were briefed by numerous impressive leaders—among them both operators and top-level staff—who were clearly providing a strong driving force for transformation, but who were within a short period, sometimes less than 1 year, of either retiring from the service or moving on to the next assignment.

This lack of continuity presents a major challenge for the military. Because the DOD is a government organization, its senior leadership is expected to rotate on a regular basis. The average tenure of a secretary of defense is 18 months, and while senior military leaders are expected to remain somewhat longer, both tenures are short compared to the time needed to effect major cultural changes.[5] Thus, sustaining attention to large issues such as interoperability requires the existence of an institutional process to facilitate such change, rather than relying on a strong personality.

Because DOD is not constructed in such a way that a single individual personality can readily create the focus needed for change, it must instead rely on organizational entities that persist over time. Recognizing that the functions of C4I are the constituent elements of an integrated whole and that the C4I "fabric" must be treated as a global system in order for the functions of command and control to achieve optimum performance, the DOD in early 1998 consolidated intelligence, security and information operations, C4ISR and space systems, and the Chief Information Officer's responsibilities in the office of the Assistant Secretary of Defense for C3I. However, a year before this new structure was established, DOD decided to separate intelligence oversight from the Assistant Secretary of Defense for C3I and subordinated command, control, and communications oversight inside the Under Secretary of Defense for Acquisition and Technology. The decision directing the separation also would have eliminated

---

[4]AEGIS is a ship-based combat system that is capable of simultaneous operation in anti-air, anti-surface and anti-submarine warfare modes.

[5]For example, the statutory limit for members of the Joint Chiefs of Staff is 24 months (with the possibility of renewal); the average tenure is approximately 4 years. The average tenure for other senior leaders is approximately 2 to 4 years.

enforcement authority, leaving the residual bodies as policy-making organs.[6]

Finally, an institutional sense of urgency is also needed to create revolutionary change. Within the DOD, actual conflict and operational deployments are often required to move the system faster.[7] Sometimes, a large, looming threat creates the same urgency, as the Soviet threat in the late 1950s and early 1960s drove the U.S. strategic nuclear program. But absent immediate conflict or looming threat, it is difficult to motivate rapid change. Moreover, lack of urgency in the DOD today is also underscored by the apparent unwillingness of the system to follow through on identified opportunities with rapid fielding, follow-through that would require significant reallocation of resources between weapons systems and C4I systems. In short, DOD does not exhibit persistent leadership in this area today.

4.1.4 Process Reengineering

Experience in the private sector with the application of information technology suggests that modest improvements are possible when such technology is used to automate existing processes. Applying information technology for such purposes is relatively straightforward, and most organizations are capable of using information technology in such ways to achieve incrementally faster and more accurate information flows and more efficient business processes.

The private sector has often found that radical (rather than incremental) improvements leading to real competitive advantage can be achieved only by significant reengineering of processes, operating methodologies, and organizations to exploit fully the capabilities enabled by information

---

[6]The committee's view of this separation is negative. Those who viewed the separation as positive sometimes argued that C3 (rather than intelligence) is the "glue" that holds most weapon systems together, and thus that C3 should be institutionally integrated with weapon systems acquisition more than with intelligence. In this view, intelligence is regarded as a less real-time function that serves the political leadership in prioritization as much or more than it does the warfighter, while surveillance and reconnaissance are part of C3/weapons systems. The committee respects this argument but believes that the weapons system acquisition culture is so fundamentally different from C4I and so much more dominant within DOD that integrating the two would inevitably result in C4I being treated like weapons systems—a fundamentally misguided treatment when the underlying technologies are so different. Furthermore, the committee believes that the future of intelligence on the battlefield is that it must become more real-time in nature to be more useful to the warfighter.

[7]For example, the GBU-28 "Bunker Buster" bomb was assembled in record time to support targeting of hardened Iraqi command bunkers in the Gulf War. The U.S. Air Force asked industry for ideas on how to destroy such bunkers in the week after combat operations started. The first operational bombs were delivered to the Gulf theater in less than a month (from project go-ahead). See the Federation of American Scientists' home page online at <http://www.fas.org/man/dod-101/sys/smart/gbu-28.htm>.

technology. Indeed, in some cases, such reengineering has resulted in the creation of entirely new business processes that are the foundation of entirely new lines of business. Such new enterprises in essence redefine the terms of competition.

Wal-Mart, a company studied by DOD as an example of the achievement of major competitive advantage, did not simply replace paper with computers in its key processes. It reengineered most of its key processes in its distribution network around the new capabilities offered by the progress of technology, dramatically improving key aspects of its business from sourcing logistics to distribution to store operations. (Note also that Wal-Mart did not rely on state-of-the-art information technology and was thus able to minimize its expenses for new technology acquisitions.) Federal Express did not simply computerize what United Parcel Service had been doing, nor did Amazon.com merely automate what Barnes and Noble was doing. They reengineered their processes and aimed for dramatic rather than incremental improvement. It was as much success in this reengineering of process and organization as the application of technology that provided the competitive advantage in each of these cases.

In the DOD context, an operational focus on how C4I can lead to improved outcomes (rather than just providing new capabilities) raises the important question of how to reengineer operational processes and procedures to achieve improved outcomes with advanced information technology. Such reengineering will take on greater urgency as new digitized weapon systems are fielded. An all-digital capability will allow information available on individual weapon platforms to be shared simultaneously and acted on from across the battlespace. Targets acquired by sensors in aircraft, for example, can be seen concurrently at multiple echelons of command and can be engaged with minimal delay by designated air or surface-based weapons operating within preestablished rules.

Reengineering can also have an impact that ripples throughout an entire organization. For example, if combat forces can be applied quickly at the right location at the right time (perhaps as the result of using C4I effectively), then there is less need for larger force structures to be prelocated to cover the range of possibilities. With smaller force structures, the need for high-volume platform modernization is diminished, the need for supporting logistics is lessened, the need for lift is lowered, the need for infrastructures is cut, and the time needed to move forces is reduced. Military doctrine can focus less on forward basing and more on rapid deployment. In short, reengineered technology-exploiting processes are likely to enable major competitive advantage for the DOD, just as they do on the civilian side.

In its site visits, the committee did see some efforts that embodied the concepts of reengineering, such as the Army's Force XXI program. Such efforts call for both business processes and combat doctrine to be reengineered. (An example of business process reengineering is the idea of reducing by a factor of 10 the personnel needed to operate a tactical operations center. An example of doctrinal reengineering is the idea of engaging in coordinated strikes across the entire 200-km-deep battlespace rather than engaging in attrition warfare.) However, in a number of cases it appeared that too little of this type of effort was under way, and that the result may be a requirement for so much technology and so much skilled manpower that the technology-enhanced versions of military units may be unaffordable, insufficiently nimble of movement, or otherwise unachievable.

4.1.5 Budgets and Reprioritization of Investment

Leveraging information technology to create large-scale institutional change usually requires the commitment of significant resources to that effort. In a world of finite budgets, such commitment inevitably entails the reprioritization and reallocation of budget lines. Moreover, given the pace at which information technology changes, ways of using information technology that are optimal today will inevitably be different in 5 or 10 years. The optimal balance and manner of use at any point in time will not be optimal—or anywhere near optimal—5 years later. Large corporations deal with such change by replacing information technology on a relatively frequent basis (i.e., more rapidly than they replace other capital investments). In the military context, balance and investment trade-offs arise at two fundamental levels: among C4I programs and capabilities and between C4I programs and weapons/platforms.

In observing DOD efforts in this area, the committee found little evidence that the very powerful statement of Joint Vision 2010 (or its service derivatives) has led to significant consequent reprioritization of resources and budgets. Indeed, because defense budget programming is undertaken incrementally, the trade-off is usually captured in terms of a question such as, Should an incremental dollar be spent on C4I or on weapons systems? This trade-off reflects a pervasive and very significant tension between the historical quest of military leadership for traditional weapons modernization and the call for investment in "force multipliers" such as modern C4I systems and applications. Furthermore, DOD does not have the luxury of rapid turnover in its C4I systems, as it often faces the tacit belief of its budget overseers in Congress and the Administration that C4I systems should have the same useful lifetime as do major weapons systems.

Because of the continued and anticipated rapid rate of advance of information technology, the appropriate balance between weapons systems and C4I technology will continue to shift, posing major challenges for the military services. DOD will never solve the C4I problem "once and for all" but will need to think constantly about how information technology could be changing the way the services perform their missions, and how best to optimize the allocation of resources among C4I systems, weapons modernization, and force structure.

The point in highlighting this issue is not to substantiate the need for a particular balance, rebalancing, or change in investment strategy. Rather, it is to emphasize that the question of balance, its evolution over time, and the impact on military effectiveness are critically important and warrant having a standing and continuing activity to look at broad investment trade-offs, with military effectiveness being the dominant consideration.

4.2 SPECIAL NON-TECHNICAL CHALLENGES FACED BY THE MILITARY

Realization of the full exploitation of C4I will require major changes in military operations and in the processes and culture of the military institutions themselves. Discussions with individuals from the top military and civilian leadership in DOD as well as with captains, corporals, and other operators in the field during exercises and experiments helped the committee to appreciate the enormity of the challenge.

This transformation is occurring (or trying to occur) at a time of significant reduction in resources. To those actively engaged in the process, reductions in resources will always appear to be a major aggravating factor making the transformation more difficult. Nevertheless, a number of committee members have participated in similar transitions in the commercial world, and note that while significant resource reductions are a major source of pain for those involved in the transformation, such reductions also can in fact have positive effects because they eliminate any doubt of the need for rapid change. In addition, this urgency can drive major reengineering rather than incremental progress, and thus produce a more positive result.

Still, the DOD faces many challenges that are not found in the private sector, challenges that are specific to the role, history, and culture of the military. None of these is an absolute inhibitor of the required transformation, but taken together they loom large. Success in moving forward at a sufficiently rapid pace will require awareness of these factors and conscious effort to deal with them.

4.2.1 Situational Challenges

Like most other modern institutions, the military lives and operates in, and must plan effectively in the face of, a highly uncertain world. While a few situations, as in Korea, provide reasonable planning scenarios, the military must in large part be prepared to respond rapidly and effectively to situations that are far less predictable, against a variety of potential aggressors, and with a wide range of potential coalition partners.

In addition, and in contrast to most private-sector institutions that have shown how to achieve competitive advantage based on technology, the military can only exercise and practice anticipated types of operations, rather than build on continuous experience, which allows incremental progress. Commercial organizations are engaged in their regular business every day, and the partners with whom they work and their competitors are a relatively stable set. Thus, real operational successes and failures are apparent if management knows how to look for them, giving decision makers a near-real-time window into the operational effectiveness of the organization.

By contrast, the competitive arena for the military is not nearly so orderly or well defined as for the private sector, and the analogy to the private sector has many limitations. While generating profit is the clear and unambiguous objective for private-sector firms, success or failure of the DOD is not something determined in the "marketplace"—as a matter of national policy, it is unacceptable for the DOD to fail. Furthermore, unlike private sector firms that practice their particular business every day, the DOD must be prepared for a very wide range of possible military operational scenarios under the constraint that (thankfully) the nation is not continually engaged in those scenarios. The military services train regularly, but the stakes involved in exercises are simply not the same as those associated with war, nor is the degree of unpredictability the same. Furthermore, live exercises are expensive. The result is that DOD must rely on a variety of surrogate indicators (e.g., the outcome of simulations, the judgments of experts) to assess itself. This set of differences is compounded by the major shift in command structure, from a service-based preparedness mode to a joint task force operational mode, which occurs upon deployment.

4.2.2 Organizational Challenges

By law, the services have the responsibility to organize, train, and equip their forces. As such, they control the budgets for their acquisition programs. Service program managers—who are responsible to the service—will naturally pay greatest attention to satisfying program requirements that are most desired by the service. For those situations in which interoperability is both not a service priority (for whatever reason) and also entails additional expense, budget pressures work against interoperability. (Box 4.1 provides an illustration of how military culture, the acquisition system, and doctrine can affect system design for data com-

BOX 4.1 Data Sharing, Acquisition, and Doctrinal Reengineering

A good illustration of how cultural and institutional factors can affect system design involves the issue of data interoperability. As discussed in Chapter 2, data incompatibilities between systems are a major source of interoperability difficulties. But in addition to technical reasons that lead to data incompatibilities, cultural factors can also create significant impediments to reaching agreement on data definitions.

Many of these factors manifest themselves as operational and doctrinal concerns. One concern is that delivery of information to a subordinate may be confused with authority to act on that information. Traditionally, flows of information to lower echelons have been limited by available resources and accompanied by direction (commands). But because interoperable systems by definition facilitate information exchanges with few constraints, they make it much easier to separate information from command. As a result, doctrine based on combining information with command or authority is threatened by a reengineering of doctrine that involves the separation of information from authority.

One illustration of this threat to existing doctrine is that a common operating picture shared at all levels of the command hierarchy enables a mode of command known as "command by negation." In this mode, subordinate units—possessing a clear picture of the overall battlefield and knowing the commander's overall intent—act on their own initiative consistent with the commander's intent. Thus, they need not wait for approval or direction from higher authority and can operate at a much higher tempo. If the commander observes the unit doing something inappropriate, or if the commander's intent changes for the unit's operating area of responsibility, he can direct the unit to do something else.

A second illustration relates to trust among units from different services and even from the same service. Data collected organically by a unit (locally collected data) are often regarded by that unit as more trustworthy than data fed to that unit from other sources (foreign data). The reason is that the meaning and value attached to data are contingent on the circumstances under which the data were collected and on the theory used to collect the data, and the unit is much more familiar with the theory and circumstances surrounding locally collected data (and thus its limitations and qualifications) as compared to foreign data. Handling of data that a subsequent system will regard as foreign data is considered risky, because the performance of the subsequent system or operational element becomes vulnerable to imperfections in that data. Because it is the characteristic of interoperability that enables data to be passed for subsequent use in the first place, interoperability can be regarded as a kind of threat to system performance.

continues

> Expectations of trust also affect the timing of information exchanges. A unit may prefer to delay the sharing of data because the data might not be fully analyzed or complete. However, users (e.g., higher-level commanders) may in many circumstances find incomplete and partially analyzed data delivered in a timely manner to be more useful than more complete and more fully analyzed data delivered late. Interoperable C4I systems enable sharing to take place at a time and under circumstances that are less controllable by the unit providing the data.
>
> These matters of trust help to explain the sense of proprietary ownership over data that the committee observed from time to time. "I want the raw data" is often heard as a rallying cry. Data (as opposed to information, which is processed data and/or fused data) is thought to be the absolute truth, and those who control data often are regarded as having substantial decision and resource leverage. In a truly integrated operation, of course, data belongs to everyone. However, the committee was frequently exposed to controversies about the view that "the commander of the Joint Task Force owns the data."
>
> Operating and doctrinal factors such as those described above are often reflected in the acquisition process. The acquisition system is supposed to reflect the concerns of users, and thus it is probably not accidental that there are no mission area requirements for sharing information, and existing documents are inadequate. The mission need statements for C4I systems generally provide no guidance against which a program can be tested, and state only that "the system must be interoperable." The concept of operations does not usually specify the information flows, and thus cannot explicate the value added at each node of a C4I network or why various information flows are important.
>
> Finally, data sharing may well have cost implications. For example, the Theater High Altitude Area Defense radar system was designed to provide track data. But if it is also called on to provide imaging information, a new processor could be required for the system. Moreover, if some other military operator were to ask for Theater High Altitude Area Defense radar data for use in some other context, he would likely meet with stiff resistance from the system owner, who would probably be concerned about repercussions in the event of misuse of the data or might worry about inadequacies in the requesting authority's processing or interpretation methods.

patibility.) Furthermore, even when a C4I program of one service is required to accommodate interoperability needs originating from another service, the new requirements for such functionality are often not accompanied by additional budget authority. There is thus no economic incentive for the other service to restrain its wish list.

The services also have the responsibility for training the forces. A major component of training is the unit exercise. Because unit training is frequent, C4I systems from the same service must interoperate often, with the result that interoperability problems are more likely to be identified. In this context, the units have substantial "local" incentives to fix these problems, because left unfixed, they will recur frequently. By contrast, joint exercises and training are relatively infrequent and involve a set of variable C4I interactions among the units that happen to train together. Thus, the immediate pressure to fix problems arising in a joint context is considerably less compared to the pressure to fix problems that will arise in the relatively near term with a unit exercise. Local incentives are thus missing, and many interoperability problems may remain hidden because the systems are not exercised often or thoroughly enough.[8] (These points support the recommendations in Chapter 2 for more frequent and systematic testing.)

A related training point is that time-phased procurements of C4I systems create doctrinal problems. A contingency might call for Unit A, equipped with and trained in the use of a new C4I system, to work together with Unit B, which lacks the new C4I system. Unit A must thus be "backward compatible" from a doctrinal standpoint, and must be able to adjust its procedures and tactics accordingly to work effectively with Unit B. For example, the Division 21 experiment at Fort Hood promises to

---

[8]This not to say that joint exercises are always effective in helping military commanders and staffs understand the value of using C4I or in uncovering interoperability and other C4I problems. One reason is that the simulations used to represent military operations are often incapable of directly interfacing with real-world operational C4I systems. Therefore surrogate systems are created to carry out the exercise. Because much of the utility of C4I depends on what seem like small details, a lack of high fidelity between the surrogates and the real-world systems may well fail to provide adequate training and to uncover interoperability problems.

The next generation of training simulations will be based on the Joint Simulation System, a system being designed specifically to have a direct interface with operational C4I systems. As the Joint Simulation System comes online, it may afford richer opportunities for joint training to incorporate effective use of C4I and to uncover interoperability problems that exist today. Nonetheless, simulations will always fall short, because the unimportant "details" of two systems that may not be captured in a simulation are often the very causes of subtle interoperability problems.

More radically, proposals have been floated by the Joint Staff to create standing joint task forces (JTFs) with 3-star leaders, thus creating organizations that can exercise jointly on a regular basis and so shake out problems in operating jointly. Thus, when a crisis occurred, the standing JTF would be ready to go, except for special capability packages that might be needed in a particular contingency. Of course, a standing JTF has value only to the extent that a "core" JTF can be defined that can handle a wide range of contingencies with a relatively small addition of special capabilities. Whether such a core JTF can be defined remains to be seen. A model of standing organizations is the way the Navy deploys battle groups to sea—the group trains continuously, and if a crisis occurs that demands a naval response, the deployed group is in position to handle it.

produce one or more divisions equipped with fully digital C4ISR systems and services by the beginning of the next decade. Unfortunately, the majority of the Army will not have access to such capabilities until years later, unless substantial budgetary change is enabled by Congress.

4.2.3 Schedule and Budget Challenges

Schedule and budgets put tremendous pressure on military C4I acquisitions. One manifestation of these pressures is the importance of reprioritization in light of the leverage of information technology, as discussed in section 4.1.5. In particular, the trade-offs between system functionality and interoperability or security can lead to significant reductions in interoperability or security in order to meet schedule and budget commitments. As a result, directives intended to assure jointness and interoperability of C4I systems have proven relatively ineffective because program managers and the services have few institutional incentives to comply with them, and few penalties accrue to C4I programs that are not interoperable. For example, despite an Assistant Secretary of Defense for C3I directive making the Joint Technical Architecture mandatory for all C4I systems, a DOD Inspector General report found non-compliance in the plans of a large number of C4I programs.[9] If such non-compliance is found in a program's written plans, it can only be assumed that some others that have compliance written into their plans will not in fact comply. On the security side, there are clear operational trade-offs of system assurance and security against connectivity, functionality, and convenience of operation. Effective systems assurance consumes money and management energy without the outward appearance of providing additional functionality. In fact, an assured system does provide more functionality. It just takes active hostilities to show it. Also, system assurance is made more difficult by the interconnection of large networks of systems that are key to realizing the C4I vision.

A related budget point is that in the private sector, attempts at integrating disparate information systems are generally accompanied by significant budget authority that is controlled by management responsible for the integration effort. In some cases, budgets of the programs to be integrated are even taxed to produce the centrally managed integration budget, which then gives those programs strong incentives to work constructively with the integration authority. A similar situation does not apply to DOD. While certain C4I oversight offices within DOD do have the ability to withhold budget authority from the services for C4I pro-

---

[9]DOD Inspector General Audit Report. 1997. *Implementation of the DOD Joint Technical Architecture* (see Report No. 98-023), Department of Defense, Washington, D.C.

grams that are not paying sufficient attention to C4I interoperability, they do not in general have budgets of their own to spend on efforts to promote interoperability. Stopping programs that do not comply with requirements for interoperability requires identifying them in the first place, and then investing time and political capital—a highly inefficient process.[10]

A final budget point is that because individual programs are often funded on a line-item basis by the U.S. Congress, program managers must explicitly account for spending funds designated for one program on another. While this requirement promotes accountability of taxpayer funds, it does pose a potential impediment when unforeseen expenditures may be necessary in one C4I program to support interoperability with other systems. Logic may well dictate that reprogramming funds from one C4I program to another is the most effective means to achieve interoperability, but such decisions may be questioned when the reprogramming report is made. To the extent that such reprogramming is made difficult, it may well be harder for program managers to make the trade-offs necessary to achieve integration.

4.2.4 Coalition Challenges

The necessary and correct working assumption that all future military operations will be joint and will most likely involve coalition partners places a special set of challenges on achieving true interoperability and security among C4I systems among all parties in such operations. There are many challenges to be faced in that regard:

- *Inability to fully predict coalition partners.* While all future U.S. military operations can reasonably be expected to employ the four military services, the coalition partners are generally unpredictable and subject to continuous change. This greatly complicates (indeed, renders essentially impossible) any effort to plan for C4I interoperability in advance of the formation of a coalition.
- *Inadequate investments, incompatible architectures.* U.S. military budgets are large compared to those of other nations on an absolute basis. Potential coalition partners, for the most part, lack adequate resources to modernize their C4I systems, and thus may well be using equipment that

---

[10]The ability of an oversight office to stop such a program depends in large measure on its ability to judge the program against some criteria for assessing interoperability. In the absence of such criteria, it will be very difficult to stop a program. And, for C4I systems, such criteria depend on the existence of operational architectures or their equivalent that specify information flows and systems architectures that specify interconnections.

is substantially incompatible with present and planned U.S. C4I systems. For example, the committee viewed a South Korean command and control system based on 1950s 60-word-per-minute teletype technology. An exacerbating issue in this regard is that the extent to which other nations favor indigenous military procurements of C4I systems diminishes the likelihood that multinational systems will interoperate readily, pending the development of technology that might render such difficulties inconsequential. In fact even when multiple nations (e.g., NATO members) ostensibly subscribe to certain C4I interoperability standards, interoperability is far from assured (as is often the case with international commercial standards).

• *Trust and security.* The United States places many restrictions on the types of information it is willing to share with certain coalition partners. Such concerns are understandable in light of the fact that today's coalition partner may be tomorrow's adversary. However, from a technical standpoint, developing interoperable information systems that allow only selective passage of information creates major challenges.[11]

• *Doctrinal differences.* As noted in Chapter 2, U.S. military doctrine emphasizes the importance of devolving operational control to the lowest levels of command consistent with centrally determined top-level goals. This doctrine makes sense for the United States, but it is not accepted by all potential coalition partners. Differences in doctrine can lead coalition forces to misunderstand information or direction coming from U.S. forces, and vice versa.

• *Language.* Language differences are a major impediment to interoperability. While these can be managed to some degree with bilingual liaison officers, the inability to exercise command and control directly in a common tongue can cause critical operational problems.

4.3 THE ACQUISITION SYSTEM

4.3.1 Overview

Success in ensuring that competitive advantage is achieved in the C4I arena requires that the changes in the DOD environment extend to

---

[11]Note also that a large impediment to coalition command and control interoperability is the indiscriminate use of secret network communications. In many cases, neither the hardware and the software nor the data to be sent to coalition partners is classified, but for some reason (e.g., convenience at some earlier stage in the pipeline), data go onto a secure network. The classification follows the data, which are then denied to the coalition partners, even though the data are in fact not sensitive. (Alternatively, manual intervention is required to remove the NOFORN markings indicating "no foreign distribution," and there are no people to spare to perform the intervention.)

its mind-set and processes, and to the regulatory and oversight structure, for acquiring new technology. The organization, procedures, and regulations governing acquisition of military capabilities are oriented largely toward major weapon systems for which the time from concept definition to fielding of the first article of production typically ranges from 10 to 15 years. This process is designed around a series of phases and checkpoints, beginning with the development of a validated mission need and progressing through concept definition, program formulation and risk reduction, engineering and manufacturing, and production, deployment, operations, and support. Further, there have historically been tensions between the DOD and the commercial providers of weapon and C4I systems, tensions involving budgetary considerations, compliance issues regarding contractor and acquisition authority relationships, and increasingly inquisitive, often hostile, news media. These tensions have typically driven significant increases in the acquisition program delivery time. As program length has increased, the pace of evolution of electronic system technology—particularly information technology—has also accelerated, now averaging a factor-of-10 improvement in capability every 5 years. Thus the standard acquisition process ensures that no program has a reasonable possibility of delivering current C4I technology to the warfighter.

The DOD no longer enjoys the leverage it used to exercise regarding the development and application of technology. The government cannot compete successfully with industry for the intellectual resources needed to satisfy its requirements, any more than it can hope to employ traditional acquisition methods and controls to leverage technology. The market drives the directions of research, engineering, and technology resources, and the market is a reward- and incentive-driven environment. The most likely means for success in leveraging these factors to the advantage of the DOD is for the department to pursue a pattern of behavior consistent with the forces of the market, that is, to participate in the market as a *consumer and partner*, keeping pace with market developments and providing capital incentives as its means of leveraging commercial system solutions to its requirements.

The leaders of the acquisition process must also face the realities of a diminishing force structure even as requirements for military capability grow. An important management challenge to DOD leadership is to achieve timely provisioning of military capabilities that can produce a higher likelihood of success of military operations, over a more varied spectrum of tasks, with fewer resources. In the context of a smaller force structure, it is worth noting that even as the military strength of the nation, in terms of actual personnel strength, has shrunk by more than 40% over the past 8 years, the oversight of the acquisition process has not decreased in proportion to that reduction.[12] The current requirements generation, acquisition, and program management culture of the DOD requires significant alteration in order to capitalize on the immense changes that information technology has brought to industry and the world. While numerous DOD leaders appear to espouse the tenets of "acquisition reform," the behavior of program directors and managers has evolved little—nor has that of the huge oversight bureaucracy instituted to ensure that every acquisition of significance satisfies the traditional acquisition regulations.

The willingness to take risks and the use of effective risk management are essential to realizing the full potential of C4I-enabling technologies. Traditional risk-avoidance strategies invariably cede the advantage to smaller and more agile enterprises, even when the "right" decisions are made.[13] In the commercial sector, the entire history of information technology suggests that it is very difficult to predict what the important information technology applications will be 3 years (roughly two current generations) in the future. Many information technology applications and capabilities are proposed; a few survive, while the majority fail. Marketplace success (i.e., large market share) often goes to vendors that deploy an initial version of a product with minimal functionality and then go on to improve and upgrade the product after fielding it.

One key area in which greater risks must be taken is in taking advantage of the flexibility already built into the acquisition system. The existing acquisition process was redesigned to (in principle) allow considerable flexibility in the program management, but that flexibility is seldom put to use by program managers. In practice, conservative "by the book" approaches better suited to long-lived weapons systems are still preferred, even for C4I programs. The training that acquisition managers receive does not prepare them well to understand the intrinsic differences between C4I (information) and weapon systems, and they may be ill-prepared to argue the significance of those differences before acquisition boards and oversight councils.

While significant modifications to some processes have been made to deal with this issue, much more needs to be done. The committee sees the need for changes in the several areas that are described in section 4.3.2 through section 4.3.6.

---

[12]Indeed, one can argue that personnel devoted to acquisition oversight should correspond to the size of the acquisition program, which has decreased by considerably more than 40%.

[13]Clayton M. Christensen. 1997. *The Innovator's Dilemma: When Technologies Cause Great Firms to Fail*, Harvard Business School Press, Cambridge, Mass.

4.3.2 Requirements, the 80% Solution, and Functional Specifications

The realization that the rate of change in technology as well as in operational requirements (especially in C4I) is not matched to the typical multiyear cycle time for traditional system acquisition has led to the concept of evolutionary acquisition, also known as "spiral development." Given a validated requirement and an approved architectural framework for future development, evolutionary acquisition allows more rapid deployment of systems and provides a process for incremental upgrading of fielded systems. Conceptually, the requirements, definition, testing, and fielding steps of traditional acquisitions are executed over much shorter cycle times for each incremental deployment. Evolutionary acquisition permits incremental addition of capabilities to a system and the underlying technologies evolve without this being viewed as "requirements creep."

A fundamental tenet of evolutionary acquisition is acceptance of the "80% solution." Insistence on a "100% solution" can radically increase costs and extensively delay system deployment. It should be stipulated that an "80% solution" is the goal for virtually all C4I acquisitions. The rationale is simple: no C4I system requirement can be effectively specified to the 100% level, nor can any C4I acquisition program deliver a "final" solution. An 80% solution allows the program design to take advantage of the inevitable changes in the underlying information technologies. It also provides a base of experience on which to specify and build the remaining functionality. And, it allows a more gradual path for possible changes in doctrine and tactics for using the capabilities provided by a new C4I system.

A good C4I acquisition, particularly an evolutionary acquisition, actively engages the end user in the acquisition cycle, particularly from the concept development through the design stage, and again in the test and acceptance phase, as noted in section 4.3.4. The end user should have significant influence in determining the "80%" point in the acquisition contract, and should be able to interface freely with the contractor (in concert with the acquisition program manager) regarding technical, cost, and capability trade-offs during the engineering and integration stages of the program. Rapid prototyping and similar techniques are useful ways of capturing user input quickly in the concept development and design stages.

A prime example of successful "80% rule" application is the Global Command and Control System (it is one of a very few major C4I acquisition success stories). The Global Command and Control System objective was functional: replacement of the antiquated World Wide Military Command and Control System with new, high-technology-based global C4I system capabilities—without sacrificing the essential capabilities of the legacy system.

The Global Command and Control System was excluded from traditional acquisition oversight, but only because it avoided designation as a major acquisition program. It was not, therefore, initially subject to reviews by the Major Automated Information Systems Review Council, nor was it required to pass through the test and evaluation process prescribed for "weapon system" acquisitions (a term that appears applicable to all acquisitions of sufficient value to warrant a Major Automated Information Systems Review Council review, as well as others—determined by the judgment of various officials in the acquisition process chain). As a consequence, the Global Command and Control System effort succeeded in achieving a very rapid (in about 2 years) replacement of the World Wide Military Command and Control System with state-of-the-art technology and a modern architectural construct that facilitates insertion of new technology as it becomes available.

This required a departure from the acquisition mind-set requiring formally validated specifications as requirements. Since the pace at which information technology advances drives the rate at which it must be exploited, one must be willing and able to accept and manage the risks attendant with reduced oversight from the acquisition community. The Global Command and Control System evolutionary acquisition process was, nonetheless, loosely based on the traditional acquisition model; but it was more rapid and flexible. The phases and milestones were set much shorter, consistent with a 6- to 18-month Global Command and Control System implementation schedule. Following rapid completion of the equivalent of milestone II, the program developed through repeated evolutionary cycles as new requirements became known. The Assistant Secretary of Defense for C3I-approved model contained six steps for each review phase: (1) identify requirements, (2) validate requirements, (3) assessment I, (4) prioritize requirements, (5) assessment II, and (6) develop. The steps were tailored to program needs, and decisions were delegated to as low a level as possible.[14]

A hybrid of traditional acquisition and evolutionary acquisition is also possible, as illustrated by the acquisition practices of the Special Operations Command. This approach initially takes a program through a normal DOD-5000 acquisition cycle, including all milestones, but then at milestone III (deployment) switches to evolutionary acquisition to incorporate

---

[14]For particulars on Global Command and Control System methods vis-à-vis evolutionary acquisition, see Richard H. White, David R. Graham, and Jonathan A. Wallis. 1997. *An Evolutionary Acquisition Strategy for the Global Command and Control System (GCCS)*, Institute for Defense Analyses Paper P-3315, IDA, Alexandria, Va., September.

mature technologies or to infuse research and development products where absolutely necessary.

The effectiveness of a particular C4I system must be judged by its ability to perform specific desired functions within a defined range of scenarios. The set of scenarios must be well chosen, as they will determine the necessary types and levels of interoperability among subsystems and systems. The implications of these choices flow through the operational and systems architectures to systems design. In most cases system capabilities will be planned for improvement over the system life cycle, building on the progress of the underlying technologies. Hence, system design must begin with the operational architecture of the system. An optimal set of evolutionary implementations will embody trade-offs between desired functionality and performance on the one hand, and what the expected progress of the key technologies will allow to be rapidly and cost-effectively deployed on the other.

Requirements for C4I systems should specify overall system functionality and performance, as opposed to detailed design specifications that are typical of weapons systems. There are always multiple solutions to a C4I requirement, and the trade-off between communications, processing, human-machine interfaces, system architecture, etc., depending on how addressed, can result in markedly different approaches and capabilities. The commercial market is the development driver for the technology pertaining to C4I; hence, military specifications to the level of detail customary in historical system acquisitions can result in unfavorable cost and capability impacts (as opposed to letting the bidding vendors propose the best solution for a functional requirement).[15]

4.3.3 Exploiting Commercial Technology

Military C4I systems depend on two very different classes of technology. One class of technology has historically been dominated by government needs (including those of the military). In the C4I domain, these technologies include sensor technology and, often, hardened communications infrastructure, and government is the primary customer as well as the principal funder of their development. Accordingly, the needs of the private sector do not affect the course of this class of technology very much. The second class of technologies is the set of information process-

---

[15]Note that it is not unreasonable for a functional requirement to include a statement of compatibility or interoperability with other systems. As discussed in Chapter 2, the easiest way to facilitate interoperability may be for the system in question to conform to a specific architecture that is common to a number of systems. For this reason, a functional requirement that specifies an architecture is not necessarily a contradiction in terms.

ing, communication, and decision support technologies whose primary market is in the private sector, and whose pace of development and detailed properties are dominated by the forces of the marketplace. This class of technologies is frequently referred to as commercial off-the-shelf technology, or COTS.

COTS technology is a "playing field leveler" of unprecedented proportions. It cannot be controlled or limited by government regulation, commercial interest, or military preference. The technological leverage the government enjoyed in the post-World War II era has disappeared in all but unique applications, such as specialty weapons and extremely high-survivability C4I systems. The course of technological advancement is now determined in world markets primarily by the ultimate bill payers, that is, the end users of the tools, processes, and applications enabled thereby.

Since the power of emerging technology is no longer under government control, and the government cannot solve its firepower, interoperability, and other technology shortfalls through specification, mandate, or other historical leveraging methods, it must learn to behave like a consumer. For reasons of economy and speed of acquisition, DOD will have to take increasing advantage of commercial technology, with as little change thereto as possible, for all digital information, research, and operational needs. DOD, like most public and private enterprises, must learn to adapt much faster to the forces of technology advancement and bring the resultant new tools and capabilities into its inventory of operational systems as they become available, not after they are obsolete. Doing so requires people familiar with digital-age technology and also demands major changes in the regulatory, operational, and doctrinal structures of the DOD and all its constituent parts.

Comparing commercial interests to the traditional view of the government as a customer leads to a far different industry perspective than that of only a decade ago. The increasing rate of advancement in technology, particularly in computers, software, and data transfer media, has created a "churn" rate unparalleled in business history. Businesses spring up and disappear or are absorbed by other businesses at a dizzying rate. Any business, large or small, that does not learn to adapt very quickly to the reality of "real-time" consumer expectations and fast-paced technology advancement is at very high risk.[16] Every information-based business today is in fierce competition with all others, not only for market share and revenues but also for limited engineering, networking, security, and software expertise.

---

[16]Regis McKenna. 1997. *Real Time: Preparing for the Age of the Never Satisfied Customer*, Harvard Business School Press, Boston, Mass.

Designing, acquiring, fielding, and upgrading complex C4I systems in a fashion that fully exploits the rapid progress of commercial technology is a significant challenge. Moreover, the challenge is usually amplified by the necessity for simultaneously providing a high degree of security and robustness against information warfare attacks. COTS components are developed and brought to market with little attention to security, because for the most part security considerations do not enter into the development of most general-purpose COTS products, as noted in Chapter 3. Thus, military C4I systems built out of COTS components must take special account of potential security weaknesses. For example, the fact that a system has potential weaknesses makes it particularly important for system designers to specify essential and non-essential functionality in the event that security is breached, to plan for failure (e.g., backups and alternative paths to achieve critical functionality).

Riding the wave of commercial technology will be difficult for the military, and it requires a high degree of technical and system competence, diligence in anticipating and tracking market advances, and ability to envision applications for those advances faster than adversaries can. Speed of application becomes ultra-important when technology is equally available to potential enemies.

DOD has made significant—but not altogether successful—attempts to move in the direction of exploiting commercial technology. In the recent past, the services' Persian Gulf experiences and those of the defense-wide Defense Management Review in the early 1990s combined to drive a new look at the possibility of exploiting commercial technology in C4I systems. The operators began calling for COTS technology in requirements statements for new information systems. This drove significant modification in the views of many program managers, but was more successful in creating expectations than in satisfying them. The goal of using commercial technology in military systems without modification has not been generally realizable. In some instances, such as software procurements for desktop applications, the shrink-wrapped versions of software have proven satisfactory; however, commercial products intended for consumer, business, or industrial markets generally require adaptation for application to operational military requirements.[17]

Interoperability remains a problem for commercial information technology, especially in the software domain. DOD wants "interoperability"

---

[17]For example, the cryptographic interfaces for COTS communications gear and macro programming languages for standard tools (e.g., spreadsheets, documents) can play an important role in building a military application out of COTS components. This phenomenon—the need to adapt COTS products for DOD requirements—is an implication of the 80% rule at work, and should not be surprising. Nor should it be used as a rationale for avoiding the use of COTS products.

between its myriad electronic systems, and values "open systems" as the enabler for achieving that objective. Industry wants freedom to develop proprietary solutions to the demands of the market, and regards "open systems" as being in tension with the protection of proprietary rights. DOD often attempts to assure industry that "open systems" and proprietary rights are not mutually exclusive, since the desired level of "openness" is limited to standard, open interfaces between modules and subsystems and does not pertain to the innards of software or hardware where the functionality of systems resides. Industry often does not believe such claims. DOD is, after all, interested in multisource form, fit, and function replacements for modules and subsystems, in order to keep prices under control, and the easiest and most effective way to procure form, fit, and function replacements is for DOD to furnish potential suppliers with the designs and source code associated with the desired modules or subsystems.

Intellectual property rights also figure strongly in another respect: that of DOD as a source of technology risk investment. DOD is regarded as having an appetite for ownership of intellectual property developed under government contracts, with an eye toward turning it over to a contractor's competitors in order to create multisource procurement potential. This is unacceptable to industry in a world where intellectual property is regarded as the most important factor for survival against highly agile, fast-moving competition.

While it must operate with the knowledge that, for many technologies, it is just one consumer in a vast market, there are a number of ways in which the DOD has some potential to influence the direction of commercial technology to better meet military requirements:

- *Participating in standards efforts.* For example, TCP/IP was not designed for operation in a mobile environment—a capability that would be very useful to DOD. DOD has participated in standards-developing forums to incorporate such desired features into future releases of the TCP/IP standard.[18]

- *Funding the development or deployment of required technologies that may have later commercial application, with the understanding that resulting products will be made available to DOD as supported "shrink-wrapped" products.* For example, the National Security Agency recognized the growing dependence of DOD on commercial communications satellites in the early 1980s. Classified space-qualified cryptographic modules were provided

---

[18]A recommendation that the Army participate in these efforts was made in the National Research Council's Army multimedia study. See National Research Council, Board on Army Science and Technology. 1995. *Commercial Multimedia Technologies for Twenty-First Century Army Battlefields: A Technology Management Strategy*, National Academy Press, Washington, D.C., page 56.

to commercial satellite builders to be integrated into their space platforms and ground stations so that the satellite command links could be protected against unauthorized commands. In addition, some of the satellite system operators reviewed their command and control systems and redesigned them to eliminate vulnerabilities to threats from outsiders and disgruntled insiders. The driving force was that DOD would only lease commercial service on satellites launched after a set date if the satellite had a secure command link.

- *Establishing mechanisms for early warning of commercial developments.* Product prototypes typically emerge as concepts 2 to 3 years before their commercial release. This lead-time would allow DOD to intervene early to either incorporate DOD requirements or leave an opening (software changes, extra chips) in these products for DOD customization. The Defense Advanced Research Projects Agency, service laboratories, and the like are suited to take advantage of such early warning to develop modifications or add features to emerging COTS technology.

- *Investing in DOD-unique changes and additions to commercial products without losing the benefits of using commercial technology.* Since product features are increasingly a function of software, these modifications tend not to impose any additional physical constraints on the product. In some cases, however, this approach might require such measures as requesting (or paying for) empty chip sockets to be placed in COTS products.

- *Leveraging DOD's unique role as a neutral arbiter in mediating industry convergence on standards.* Competing commercial interests may block progress important to information technology developments of interest to DOD. In some instances, DOD may be able to play a leadership role in helping these varied interests to reach convergence. An example of such success was the DOD role in facilitating the President's Advanced Distributed Learning Initiative.[19] Key to this initiative was development of an industry-agreed technical architecture for the distribution of interac-

---

[19]The Advanced Distributed Learning Initiative (ADLI) is designed to "ensure access to high-quality education and training materials that can be tailored to individual learner needs and can be made available whenever and wherever they are required. This initiative is designed to accelerate large-scale development of dynamic and cost-effective learning software and to stimulate an efficient market for these products in order to meet the education and training needs of the military and the nation's workforce in the 21st century. It will do this through the development of a common technical framework for computer and net-based learning that will foster the creation of re-usable learning content as 'instructional objects.'" In order to facilitate the development of specifications that meet the interests of all participants (from both government and the private sector), the ADLI will ensure that a common set of guidelines for this new object-oriented learning environment is developed through active collaboration with the private sector, where many of the innovations in network technology and software design are taking place. For more information, see the ADLI Web site at <http://www.adlnet.org>.

tive learning software over the Internet. Given the intense competition among commercial parties, closure on this work was difficult without DOD's willingness to assert leadership as a trusted third party to produce a reasonable result.

4.3.4 Testing

Much of the emphasis in acquisition for the last 10 to 15 years has been on separating users from the acquisition process except at very precisely defined points (e.g., the mission needs statement, the operational requirements document). The separation of testers and buyers, after many years of their working together, was the result of (the perception of) many cases of slipshod and prejudiced testing. In the 1980s these functions were separated by congressional mandate, an action that will make any change in this area difficult.

The nature of C4I systems in general, and of systems developed using a spiral development approach in particular, calls for a more cooperative and collaborative approach to program testing:

- By insisting on a separation of testers from acquisition personnel, the test process relies on an "over the transom" model that takes a long time to execute.
- The process does not put systems into the hands of users early enough to allow refinements or mid-course corrections prior to fielding.
- The current model is based on the full specification of system requirements in advance. However, all of the requirements for a C4I system are not usually known in advance. For example, experimentation might reveal new ways of using the technology that were not originally anticipated. Furthermore, the time needed to meet 100% of the requirements is often too long compared to the rate of change of the underlying technology, and thus a system that must be fully specified (so that it can be tested "properly") may well be based on obsolete technology by the time it has met those requirements.
- C4I systems are hard to test in a stand-alone environment (especially for things like interoperability); they are best evaluated in real-life settings, connected to the other parts of the C4I network.

Thus it is important that end users be more closely coupled to the work that the acquisition system does—setting requirements, testing systems, and so on. The end user would be better served if the C4I acquisition process not only involved users more closely in the initial statement of component subsystem requirements and system-level requirements but also kept them fully involved in the continuing revision of those requirements to meet certain identified performance and affordability objectives

(Box 4.2). Additionally, the end users should remain closely involved during developmental testing (both at the modeling and simulation and full-scale levels) to ensure that operationally the C4I system actually performs as expected in the total system environment.

Under DOD's acquisition reforms, the roles of the program manager, the prime contractor, and supporting engineering organizations are in transition. The use of integrated product teams involving all parties above is common today.

4.3.5 Flexibility in the Process

In recognition of the need to rapidly bring new technology into the hands of users, DOD has adopted programs that take off-the-shelf subsystems and products and develop them for DOD use through pilot programs. These efforts are not classified as acquisition programs and thus are able to more rapidly and nimbly test new technologies and concepts. Advanced concept technology demonstrations (ACTDs) act as a catalyst and matchmaker to bring together mature commercial technologies and unmet user needs. ACTDs and similar programs (e.g., the Joint Warrior Interoperability Demonstrations) have also proven to be a valuable way of educating users about the potential of certain C4I applications.

However, to take full advantage of the opportunities that these programs afford, DOD needs some budgetary flexibility to exploit unanticipated advances in C4I technology that have a high payoff potential. High-value C4I applications may emerge quickly (e.g., as the result of experiments or demonstrations such as a Joint Warrior Interoperability Demonstration) or on a track other than that of a normal acquisition (e.g., as the result of an ACTD). Because ACTDs and other experiments have not been planned for within the normal planning and budget process, follow-on procurement requires both a process for insertion into the appropriate phase of the acquisition process and a means of gaining budgetary support. Because service budgets do not include extra funds for such circumstances and reprogramming funds is a difficult task (implying that the budget for an otherwise funded program must be reduced), an "off-line" funding mechanism is required to cover unanticipated needs. Finally, even if an ACTD does not enter the mainstream acquisition process, funding streams are needed to ensure that useful leave-behinds from ACTDs are kept compatible.[20]

---

[20]It should be noted that successful advanced concept technology demonstrations in all areas, not just C4I, suffer from the problem of transitioning into production. But one of the major differences between C4I prototypes and weapons or platform prototypes is that the former are often inexpensive to replicate by comparison to the latter. When this is the case, going from "technology prototype" to operational deployment can be faster and done in a more affordable manner.

BOX 4.2 On the Importance of Close Coupling to End-User Needs

When designing and implementing C4I systems, the importance of working very closely with end users cannot be overstated. From the designer's perspective, a C4I system that does useful things should be seen as a benefit. But from the user's perspective, the threshold for utility is not whether a system is "useful," but rather whether it provides more utility than other systems or ways of doing things. System designers develop systems to meet operational needs as they have been reported to them, but because these reports are inevitably incomplete (it is hard to report on all of the "little" things about a system that greatly affect its usability for and utility to the end user), designers focus on system capabilities; how the user must use the system becomes an afterthought.

In its site visits and other experience, the committee noted several examples in which the user's job was made more difficult because of the introduction of a new C4I system.

- An intelligence analyst in a battalion operations center was viewing an operating picture of the battlefield that contained information generated organically by the battalion. Even though this information did not require vetting or analysis before it was transmitted to the division operations center, manual intervention was required to update the division operating picture. A procedure was in place that required the analyst to update the picture regularly, but automation could have assumed that responsibility just as easily, with a better-synchronized operating picture being the result.
- Field commanders make many requests for intelligence information. But in fact intelligence assets are rarely, if ever, sufficient to fulfill all such requests. Higher authorities (e.g., division and corps commanders and higher) must make choices about which requests to fulfill. But the committee saw a number of instances in which a field commander did not know for many hours whether or not a particular intelligence request would be fulfilled, even after decisions had been made about allocating intelligence assets. Because a commander's operating plan may be different if he lacks the necessary intelligence information, the planning efforts of these commanders were unduly delayed. Commanders can more easily accept the fact that a particular request will be denied—because they can then plan around it—than they can accept uncertainty about its status.
- An operations officer noted that a certain C4I system previously in use provided information in printed form. Because the officer needed multiple people to use that information, he could post it on a bulletin board. However, a new system replacing the old system provided that information more rapidly, but it also did not support hard-copy display. So, in order to provide the information to all of the people that needed to see it, this of-

continues

ficer was forced to transcribe it manually onto a piece of paper that he could then post.
- When commanders send high-priority messages, they want to know that the messages have been received. In the absence of positive confirmation that a message has been received, they are often tempted to (and do!) resend messages to the same destination. Such redundant messages waste bandwidth and divert staff attention at the receiving end. System designers must thus provide for easy and convenient ways to check on the status of messages without clogging message queues on either the sending or receiving end.

These examples suggest strongly that the canonically correct approach to system design is to define (or reengineer) the user's job first, taking into account user skills, training requirements, and so on, and then design the system in order to help the user accomplish the new job, rather than focusing on system capabilities per se. Inevitably, the user-centered approach to system design requires that the designer work closely with the user—preferably side by side over an extended period of time—to uncover "little" things that a designer might not imagine on his own.

4.3.6 Support of the Legacy Base Versus New Technology

The military services have tended to retain legacy information systems that were developed in response to "stand-alone" requirements, were not regarded as subject to connection with other systems and, therefore, are not operationally friendly with their increasingly interdependent companion systems. The legacy systems issue is one of the greatest challenges faced by the DOD today. This base of information systems comprises thousands of multigeneration electronic system elements and billions of dollars of capital investment, and is kept alive through the expenditure of many more billions in support costs.

In the commercial world, such legacy systems are often kept operational based on a view their cost must be amortized before new capability can be economically justified. The military environment likewise seeks to amortize its investment; but the reasons are both functional and economic: the large-scale modernization of legacy systems entails major changes in training, doctrine, and organization, in addition to the difficulty of securing political support for new investment dollars.

4.4 PERSONNEL, KNOWLEDGE, AND PROFESSIONALISM

As C4I technology becomes more pervasive in supporting the operations of U.S. military forces, and as operational processes and procedures are reengineered, the skill set required of DOD personnel—both civilian and military—to function effectively will undergo considerable redefinition. Fortunately, the composition of today's DOD personnel is changing. Younger military people, many now moving into positions of senior rank and responsibility, are conversant and comfortable with the use of information technology tools, processes, and systems.

DOD offers its people a range of opportunities to develop expertise in C4I.[21] Nevertheless, while DOD people—especially the younger ones joining the military services—are increasingly familiar with information technology, DOD must foster and accelerate the development of a culture of stronger information technology awareness within the military forces, among both staff and combat personnel. Development of such a culture requires changes in the status and the perception of information systems personnel, including effective and meaningful rewards and incentives and increased training in C4I systems and capabilities for all military forces. It is vitally necessary to attract, retain, and employ information technologists for the operation and effective use of military information systems and combat systems alike; yet the current operational leadership culture relegates such resources to the perceived status of second-class citizenry.[22] There are marked limitations in promotion opportunities, education and training, and command and senior leadership opportunities for personnel in the military information systems and technology fields. And, DOD's efforts to retain qualified personnel in information technology are complicated by a general shortage of information technology workers

---

[21]These opportunities include offerings of the National Defense University (including its component colleges, among them the Armed Forces Staff College and the Information Resources Management College), the NATO School at SHAPE and the NATO Communications and Information School, and the Naval Postgraduate School. For more information on these institutions, see the National Defense University home page at <http://www.ndu.edu>, and especially the page for the School of Information Warfare and Strategy <http://www.ndu.edu/ndu/nducat25.html>; for the Information Resources Management College, see <http://www.ndu.edu/irmc/>; for the NATO School at SHAPE, see <http://www.vabo.cz/English/military/NATO/nat2.html> and http://www.vabo.cz/English/military/NATO/Courses/courses.html>; for the NATO Communications and Information School, see <http://www.nato.int/docu/handbook/hb32711e.htm>; for the Naval Postgraduate School, see <http://web.nps.navy.mil/~ofcinst/code39.htm>. For the most part, the C4I courses offered by these institutions are descriptive rather than technical.

[22]As an example, Navy pilots often refer contemptuously to even operational electronic warfare specialists as "geeks."

nationally.[23] Hence, the combination of industry's greater monetary rewards and opportunities for personal recognition and advancement creates a strong force beckoning to every engineer, technician, and system specialist—enlisted or officer.[24]

In order to attract and retain highly qualified information technology and systems experts, a new DOD leadership commitment is needed—to providing career paths for information technology specialists that include training and firsthand experience in combat doctrine, strategy and tactics, and employment of military forces. It is highly likely that people with information technology backgrounds will soon become operators in the strictest sense of the military definition thereof. Therefore, commitment to a vision of fully integrated and joint technical/combat forces, and the attendant opportunity to compete for and achieve command of combat units and promotion to the most senior positions of military responsibility, are necessities. With increased status also comes increased accountability for operational outcomes and greater commitment to careers in the armed forces.

A possible additional synergy between the civilian and military sectors with respect to C4I expertise is the reserve and guard personnel system. Reserve and guard personnel have "day jobs" in civilian life, and those with information technology jobs in the civilian world provide a natural coupling between that world and the DOD environment. Of course, the C4I systems that they must handle in the DOD are different from the information technology systems in the civilian world, but the

---

[23] A study undertaken in 1997 by the Defense Manpower and Data Center was commissioned to review personnel retention (among other things). This study concluded that personnel with Military Occupational Specialties related to information technology were not only declining in number, but also declining faster than the overall personnel reductions associated with the force drawdown. Such a trend does not bode well for a military that is increasingly dependent on information technology. Whether, and if so how much, the nation faces an overall shortage of information technology workers is the subject of much debate and controversy; many firms anecdotally report difficulties in hiring and retaining such workers. For more information on this controversy, see General Accounting Office, 1998, *Information Technology: Assessment of the Department of Commerce's Report on Workforce Demand and Supply*, GAO/HEHS-98-106R, March 20. The Computer Science and Telecommunications Board is leading an effort involving other units of the National Research Council to address congressional questions in this area; a report on the subject of the information technology work force is due in fall 2000.

[24] It is also worth noting that the supply of qualified information technology workers available to defense contractors is an issue that concerns DOD as well. Many C4I projects involve labor-intensive system designs and implementations and entail enormous software developments. For all practical purposes, government acquisition regulations effectively cap the salary that can be paid to workers assigned to DOD contracts. In some cases, the cap can be sufficiently low that it is impossible to compete effectively with commercial and consumer-oriented companies for the most talented information technology professionals.

systems do share certain commonalities (e.g., they are often built out of the same components). More importantly, the intellectual skills and capabilities needed to operate and support information technology systems in the civilian world are also highly useful within the DOD world.

Lastly, despite DOD's best efforts at retention, the financial lure of the private sector may well be too great for DOD to achieve low turnover in C4I personnel. If so, DOD will need to recruit personnel, train them quickly in information technology skills, and then accept that a high portion of them will leave DOD when their service obligation is completed. This trajectory, which seems quite plausible, suggests that DOD must prepare for rapid, effective, high-volume training in information technology skills. How best to provide such training is an open question at this time.

4.5 EXERCISES, EXPERIMENTS, AND DOCTRINAL CHANGE

Doctrine refers to the fundamental principles that guide the actions of military forces, i.e., how those forces fight. Doctrine is developed on the basis of the judgment and experience of senior military commanders and is promulgated throughout the services.[25] Exercises and experiments are both intimately tied to doctrine, but they have fundamentally different purposes. The purpose of an exercise is to train units to fight in accordance with established military doctrine. That is, a unit engages in exercises in order to learn how to apply to combat the principles enunciated in doctrine and to maintain readiness through training. The purpose of an experiment is to explore alternative doctrine, operational concepts, and tactics that are enabled by new technologies or required by new situations. That is, new technologies or situations may call for different ways of conducting operations. But without actual operational experience in using those technologies or in those new situations, experiments are the next best thing, because they provide more of a basis for making informed doctrinal choices than does reliance only on analytical studies and/or simulations.

Note that experiments can often be expected to "fail" (although the consequent learning is, of course, a success by definition). That is, an experiment may conducted that tests a particular doctrinal approach to using technology in a certain scenario. The results of that experiment may well show that the doctrinal approach selected has serious flaws not

---

[25]Note that doctrine may be relevant to a single service (e.g., Navy operations), in which case the formulation of doctrine is the responsibility of that service. It may be relevant to more than one service, in which case the services involved take responsibility together. Or, it may be joint doctrine, in which case it is the responsibility of the Joint Chiefs of Staff and the Joint Warfighting Center (part of the U.S. Atlantic Command effective October 1998).

apparent before the experiment. Such an outcome should be expected from time to time; the term "experiment" would not apply if this were not the case.[26] By contrast, exercises are intended to "succeed." An exercise is conducted with certain training goals in mind. Success is achieved when the units involved are able to achieve those training goals, i.e., when they have learned how to conduct themselves in accordance with established doctrine.

Achieving the vision of the future articulated in Joint Vision 2010 will require both experiment and exercise, experiment to determine new doctrine (i.e., how best to exploit information technology in support of the revolution in military affairs), and exercise to teach U.S. fighting forces the doctrinal implications for how to fight. The Office of the Secretary of Defense and Joint Staff and the military services have begun some testing, experimentation, and exercising to explore new concepts of operation and doctrine that advanced information technology makes possible. Indeed, new organizations have emerged in the services and the joint commands to explore, codify, and invest in C4I and information technology for military applications. As noted in Chapter 1 (especially Box 1.7), studies, exercises and experimentation, and recent and ongoing military operations have all demonstrated the potential for dramatic enhancement of military effectiveness through the use of improved C4I technologies and systems.

Despite current efforts, however, much more testing, evaluating, exercising, and modeling and simulation need to be done in order for the military to translate its vision of information superiority into reality. While the Joint Chiefs of Staff and the services have at least a broadly defined and understood vision of information-dominated warfare, they are just beginning to examine the more relevant issues of how force structure, weapons employment procedures, battlespace management, and logistics support can be shaped through information-enabled military dominance.

Today, the benefits and advantages achievable through aggressive use of the latest C4I technology are neither well proven across the full spectrum of potential military operations nor well understood in terms of the reengineering of operations that this technology can potentially enable. Some indications of the benefits and advantages are known from experiments and modeling, but such experimentation is not sufficiently mature to be the sole or even the primary basis for decision making regarding doctrinal changes and major trade-offs in acquisition and force structure.

---

[26]Note also that as used, the term "experiment" does not generally refer to a controlled experiment in the purely scientific sense of the term; such experiments would be too expensive and time-consuming. On the other hand, there is a school of thought that would argue that additional rigor is both practical and needed.

4.6 MANAGEMENT METRICS AND MEASURES OF MILITARY EFFECTIVENESS

Achieving large-scale cultural change in an organization requires commensurate change in management and the organizational metrics. Metrics are (or should be) important to senior decision makers in any organization. It is a long-standing axiom of quality management that "if you can't measure it, you can't improve it." In large commercial organizations, the behavior of personnel is strongly influenced by the metrics that management uses to assess performance, whether those metrics are part of a formal assessment or are more perceived than formal. People are keenly aware of what matters in terms of rewards, promotion, credit, and the like, and they behave in a manner consistent with their perceptions. Good management metrics help to drive organizational behavior that supports areas of operational significance. In general, management metrics focus on organizational performance or characteristics and are used by senior management to assess the effectiveness of the organization and its leadership. Box 4.3 lists some candidate management metrics.

BOX 4.3 Possible Management Metrics

Possible management metrics might include the following:

- Number of troops trained in the use of specific C4I systems,
- Number of C4I systems "certified" to be interoperable,
- Percentage of initial transmission messages received correctly by shooters,
- Latency of sensor information flow,
- Percentage of uptime and downtime of systems communications,
- Percentage of consistency/disparity of redundant data sources,
- Inventory reduction as a result of just-in-time management techniques,
- Number of C4I systems that conform to the Joint Technical Architecture,
- Number of tries needed to establish connections,
- Delay in sending critical command messages and time to receive acknowledge messages,
- Frequency of planning errors encountered in a planning system,
- Time or personnel required to develop time-phased force and deployment data or an air tasking order, and
- Time needed to stand up a tactical network for a joint task force.

A different class of information—what might be called measures of military effectiveness—can be used to help make decisions about resource allocation and procurement (e.g., what systems should be bought; what balance should be struck among personnel, weapons systems, and C4I). Such measures may, for example, support the case for acquiring 100 stand-off weapons rather than one attack airframe (or vice versa). In practice, measures of military effectiveness may also have operational significance for battlefield commanders, though that is not their primary purpose. Box 4.4 describes some measures of military effectiveness that could be used to better understand the impact of C4I on military operations.

Measures of military effectiveness are the variables of significance associated with the prevailing theory or doctrine of combat. By analyzing how these measures change in different combat scenarios (e.g., using different C4I systems, different tactics, different weapons), it is possible to gain insight into what combinations of tactics, weapons, and C4I systems are likely to be more or less effective for a given scenario. Undertaking such analysis over a broad range of scenarios of interest to DOD thus provides analytic support for particular approaches to investment.

4.6.1 DOD Use of Management Metrics and Measures of Military Effectiveness

The committee is aware of some areas in which management metrics have been changed in order to drive cultural change within DOD. For example, promotion to general officer rank now requires that a person must have served in a "joint" assignment. At this writing, the DOD is attempting to formulate criteria to set a standard of information security practice that could be used to hold unit commanders responsible for such practices within their command.[27] Nevertheless, it is almost certainly the case that there are additional opportunities to exploit responsiveness to management metrics in driving change.

---

[27]DOD has always imposed on its personnel requirements for maintaining appropriate security. However, although such requirements have covered practices in information security, there are at present no criteria that an individual can be said to meet or not to meet. Thus, for all practical purposes, enforcement of these requirements for information security has not been possible. (Personal communication, Captain Katharine Burton, Staff Director, Defense Information Assurance Program, March 8, 1999.) Note also that related efforts have been in the pipeline for many months. For example, public sources reported in February 1998 that the Army was preparing new computer security regulations that would outline the responsibilities that various Army personnel have for safeguarding their systems and penalties for those who are found to have failed in their duties. (See Elana Veron, "Army to Hold Commanders and Sysops Liable for Hacks," *Federal Computer Week*, February 2, 1998.)

BOX 4.4 Possible Measures of Military Effectiveness Related to C4I

1. Ongoing performance data that can be readily observed and tracked, such as:
   - Number of targets killed per unit time,
   - Number of targets killed divided by number of attempts to kill,
   - Number of targets put at risk per dollar invested in system capability,
   - Percentage of detected security penetrations thwarted per unit time,
   - Percentage of enemy attacks deflected,
   - Delay in commander's visibility of major battlefield change,
   - Decision time—measured as the delay between visibility of information and initiation of action,
   - Reaction delay—measured as the time between decision to act and completion of action execution,
   - Number of different military units that can be connected to command when needed,
   - Time between target identification and weapon-on-target,
   - Single-shot probability of kill using a given C4I system/weapon combination, and
   - Number of target engagements per unit time.

2. Observations of aperiodic failures and tallying of their root causes, such as:
   - Mishaps due to friendly fire, and
   - Erroneous battlefield descriptions.

3. Results of stimulated tests, such as:
   - Time to react to a breach of security, and
   - Time to deploy troops in response to a specific threat.

The committee recognizes that this list of possible measures of military effectiveness is not exhaustive. Further, it does not differentiate between what the Military Operations Research Society calls measures of force effectiveness that characterize how a force performs its mission (e.g., loss exchange ratios), measures of C2 effectiveness that characterize the impact of C2 systems within the operational context (e.g., ability to generate a complete, accurate, timely common operating picture of the battlespace), measures of C2 system performance that characterize the performance of internal system structure, characteristics, and behavior (e.g., timeliness or accuracy), and dimensional parameters that measure the properties or characteristics inherent in the C2 system itself (e.g., bandwidth).[1]

---

[1]For more detail on these topics, see Ricki Sweet, Morton Metersky, and Michael Sovereign, *Command and Control Evaluation Workshop* (Revised June 1986), MORS C2 MOE Workshop, Naval Postgraduate School, January, 1985; and Thomas J. Pawlowski III, et al., *C3IEW Measures of Effectiveness Workshop*, final report, Military Operations Research Society (MORS), Fort Leavenworth, Kansas, October 1993.

DOD has a long tradition of using measures of military effectiveness in a variety of contexts to help make investment trade-offs, including measures relevant to C4I. For example:

- Measures of effectiveness indicating significant imbalance in two areas of information technology were identified through analyses of C4ISR system and capability options incidental to the Quadrennial Defense Review:[28] (1) surveillance and reconnaissance capabilities are in danger of outstripping the military's ability to process and exploit the information collected therefrom; (2) data dissemination requirements, generated by these same sources and the associated C4I automation, are in danger of outstripping the military's communication capability and capacity, particularly at the tactical level.
- When the range at which a weapon can be used most effectively is constrained by the identify-friend-or-foe performance (rather than the lethal envelope of the weapon system), investment in enhanced identify-friend-or-foe capability is warranted.
- The operational utility evaluation of the Advanced Medium-Range Air-to-Air Missile system demonstrated that the improved situational awareness (360 degree coverage vs. 60 degree coverage) afforded to pilots by the Joint Tactical Information Data System would substantially enhance fighter lethality and survivability.

Other efforts under way to develop tools to assess the contribution of C4I to military effectiveness include:

- NATO Research Group (RSG)-19 is completing a "code of best practices" that characterizes the state of the art in methodologies for assessing the impact of C4I on mission effectiveness in such areas as: structuring the problem, characterizing the scenario space, formulating measures of merit, selecting and creating appropriate tools and data, executing the tools using appropriate experimental design, and deriving insights from the resulting data.
- Work is in progress on the Joint Warfare System, a new simulation tool that features enhanced representation of C4I, and on NETWARS, a new simulation providing enhanced representation of communications.

---

[28]The Quadrennial Defense Review, released by the DOD in May 1997, analyzed the threats, risks, and opportunities for U.S. national security. It reviewed all aspects of the U.S. defense strategy and program, including force structure, infrastructure, readiness, intelligence, modernization, and people. For more information, see <http://www.defenselink.mil/topstory/quad.html>.

Both are in the preliminary stages of development; several years will be required to refine, verify, and validate them.

Despite such work, the committee was told that measures of military effectiveness related to C4I are not particularly relevant in an operational sense either to units or individual commanders. With frequent updates to C4I systems owing to spiral development models and other factors, judgments of effectiveness (and subsidiary matters such as interoperability) are time-perishable, and tracking appropriate measures of military effectiveness becomes even more valuable.

4.6.2 Considerations in Assessment of C4I System Effectiveness

Experience from the private sector suggests that the benchmark for evaluating the success or failure of an information technology application should be its contribution to the end user. In the DOD context, the analogous statement is that the benchmark for evaluating the success or failure of a C4I technology application should be its contribution to the combat operator. All too often, new technology is introduced for reasons that are unrelated to end-user success except in the most indirect kind of way. For example, it is not unknown either in the private sector or in government that new technologies are introduced primarily because they simplify the jobs of those in the information systems support group.

The vision and the promise of the revolution in military affairs are that U.S. combat decision making will require much less time than that for an adversary. Thus, a reasonable quantitative measure of C4I system effectiveness is speed of command and control, including data input, analysis, and reconciliation; decision making; and subsequent action. The speed of a communications or computer system in transmitting information and the speed with which decisions can be made are both subsumed under this overall "end-to-end" measure. In addition, the speed and efficiency with which data are gathered from diverse sources, fused into useful decision support products, and distributed in a rapid, secure, and reliable manner to decision makers are all possible measures for which data could and should be compiled, suitably weighted, and used appropriately in the assessment of C4I effectiveness.

Furthermore, that assessment can and should be used to ascertain on a continuing basis the weakest links in any operational process involving C4I systems and to effect continuous improvement. In essence, speed of decision or command is a function of the decision maker or the commander. The set of decision support tools for such purposes can never be too efficient or too effective; therefore, a highly aggressive effort for continuous improvement in using C4I would no doubt have the effect of making decisions faster and of higher quality, and of making command a better-informed and more effective function.

This is not to say that speed is the only measure of command and control, or even the only measure of the effectiveness of C4I systems. In particular, the contribution of better C4I systems to the quality of decisions made by commanders is just as important (arguably more so). But where human judgments are the key to success, the value of C4I is particularly difficult to assess. After all, a sophisticated C4I system can be used to transmit incompetent orders. For this reason, the quality of decision making is mostly omitted from this discussion.

The argument above emphasizing speed of command and control does not reduce the importance of management metrics or intermediate measures of military effectiveness that have operational significance. For example, tracking DOD progress in building and sustaining joint interoperability in the field and to understand the interoperability situation is an essential component of any effort to promote interoperability. Both system managers (e.g., the Office of the Assistant Secretary of Defense for C3I, the Joint Staff Directorate of C4 Systems chaired Military Communications-Electronics Board) and operational users (e.g., theater CINCs or joint task force commanders) benefit from such tracking. But for tracking to have operational significance, assessment of C4I interoperability from the perspective of joint task force commanders and the abilities of C4I systems to support their needs is critical. Once appropriate measures are defined, suppliers of C4I systems should be asked to describe the result of their system improvements in terms of the defined measures, or to define new measures for consideration for adoption by DOD. Intermediate indicators do not directly assess the end result, but rather assess the factors that likely influence the end result positively or negatively, before the end result occurs.

4.6.3 Caveats

Management metrics and measures of military effectiveness are important components of sustaining the revolution in military affairs and characterizing the impact of the use of advanced information technology on mission effectiveness. But it is important to note several caveats.

- Both management metrics and measures of military effectiveness inform but cannot substitute for the judgment of senior military leaders. As the discussion above regarding the quality of decision making suggests, numbers aren't everything.
- Overreliance on precise quantitative evidence resulting from exercises and studies is likely to delay changes necessary to exploit the ben-

efits of C4I, because developing quantitative evidence is often quite time-consuming. Furthermore, quantitative evidence can be "cooked," and in any event may well not yield results that provide a clear basis for decisions.

• Overall measures of military effectiveness may help capture the contributions of C4I systems to outcomes, but identifying the precise contribution of C4I systems taken as a whole may well be problematic. And, it is even more difficult to identify the precise contribution of a specific C4I system. Using a given set of widely accepted measures of military effectiveness to distinguish the particular contribution of C4I to military operations may result in a confounding and confusing analysis unless proper care is taken to understand the data. (The obvious solution—to use measures of military effectiveness that are specifically tailored for evaluation of C4I—runs the risk that these measures of military effectiveness are developed for the specific purpose of showcasing and defending a particular proposed C4I acquisition.)

• A C4I system designed to meet a particular need may in fact have applications that go far beyond that particular need, and it may provide an infrastructural capability that potentially benefits a large number of weapons systems; the Global Positioning System is an example of an infrastructural technology upon which many weapon systems have come to rely. Measures of military effectiveness focusing on the system's ability to meet the initial need will not capture the broader possibilities.[29]

• Appropriate quantitative metrics or measures of military effectiveness may be very difficult to develop. In such cases, summaries based on human judgment provided in the form of "stop-light" scorecards (i.e., sets of red/yellow/green indicators) of some relevant problem area and its resulting impact on operational capability (by mission or by function) can be a useful way to measure progress. Sophistication is not particularly important; rather, the fundamental need is to move to a point where the problem area is analyzed and assessed based on considerations of operational significance, as well as facilitated in a technical sense. (Chapter 2 provides an example of how such scorecards can be used in assessing interoperability.)

A related point is that opinion surveys of users may provide useful information, if the results are tracked over time. For example, surveys of commanders' assessment of their view of the battlefield, of their ability to dynamically reallocate resources, and of their reactions to a statement like

---

[29]Similar considerations apply to the cost side. If a C4I system supports multiple weapons, it is a matter of judgment (some would say politics) as to what fraction of the cost of the C4I system should be counted for analytical purposes. (A "stovepiped" C4I system unique to the weapons system is easiest to compare, but is least useful.)

"C4I systems now interoperate adequately with each other for joint battles" provide judgment-based measures of military effectiveness that can point to problems and corrective actions.

- Because measures of military effectiveness are tied to specific doctrinal approaches to military operations, C4I systems that enable new ways of doing business (e.g., new concepts of operation or new doctrines for how to conduct military operations) are hard to assess using existing measures of military effectiveness. Indeed, a proper assessment may well require a new set of measures of military effectiveness. For example, the placement of radios inside tanks enabled the German army to develop a concept of operations—the blitzkrieg. But in the absence of clear and convincing evidence that the blitzkrieg could be a successful approach to the conduct of war (and there was no such evidence before the Germans used the blitzkrieg), it is hard to see how the combat value of placing radios in tanks would have been formally assessed against that of using additional tanks. Indeed, the German army achieved astonishing victories with relatively few casualties on both sides—measures of military effectiveness that focused on the number of Allied tanks destroyed, for example, clearly would not have demonstrated the effectiveness of the blitzkrieg.

The corollary of the proposition that new measures of military effectiveness are needed for new doctrines is that without such measures, the case for a revolution-enabling investment in information technology (or C4I) is difficult to make analytically. In the private sector, investments in information technology to facilitate or promote such fundamental change are more often made on the basis of an instinct and judgment about the inherent potential of a new concept. On balance, the result has been some remarkable successes—and many failures. Similarly, the success of the Revolution in Military Affairs will depend on the sound judgment of visionary and experienced military leaders who are open to evidence provided in exercises, experiments, studies, and simulations.

4.6.4 Ways of Generating and Developing Data

Once metrics and measures of military effectiveness are developed, a question arises as to how relevant data may be obtained. Computer simulations are one approach to investigating the worth of new concepts and technologies. Simulations can certainly provide useful information, but the information generated through many of them can be less than adequate in several ways:

- The information obtained in a simulation is not particularly vivid or memorable. Printouts and static graphics simply do not have the emotional impact of live demonstrations.

• The models underlying a simulation are usually based on an accepted understanding of current doctrine and tactics. Thus, they are ill-suited to demonstrate how a radically new doctrine enabled by C4I technology can lead to dramatically new results.

• Most models must make simplifying assumptions about the nature of combat. Thus, the fidelity of any model can always be challenged by parties opposed to the model's programmatic implications.

A complementary approach to modeling and simulation is to make use of live experiments. The DOD has embraced the concept of live experiments to a certain degree, and each service has a program of experiments to explore the use and value of C4I.[30] Live experiments have the virtues of greater realism and enable the examination of larger excursions from present doctrine and organization than is possible within the limits of a simulation. In addition, they can help to uncover a host of system integration problems, provide valuable training, and build users' confidence that they can trust C4I systems for mission success and survival in future wars.

On the other hand, live experiments are expensive to conduct on a large scale, and it is all too easy for reasons of training and economy for an experiment to explore only small deviations from the accepted wisdom, while larger deviations may be the ones that result in the largest payoff. Small-scale experiments are also inherently at odds with understanding the value of C4I applications that cut across systems, echelons, functions, and services. In small-scale settings, costs often dictate small samples and reduce the ability to control variables, and the large number of degrees of freedom makes rigorous conclusions problematic. Moreover, live "experiments" tend to attract public attention where failure can lead quickly to loss of support. Under these circumstances, the incentives are weak to structure tough tests to fully stress a system.

Two other possible ways of developing data (used in the telephone and computer industry) are the following:

• Generating periodic tests of a working system of systems and/or processes and tracking those test responses over time for trends, and

• Observing and tracking all field failures (including exercise results) and conducting a root-cause analysis on all the major ones to observe

---

[30]The term "experiment" is not quite the correct term for DOD efforts in this area. The terms "concept exploration" or "pilot demonstration" might be more appropriate, because the term "experiment" has connotations of a controlled trial, in which some variables are held constant and the impact of others on the outcome ascertained. Nevertheless, this report uses the term "experiment" in the DOD sense.

improvements or degradations and causes. Of course, such an approach presumes that appropriate instrumentation is available to capture such data as may emerge.

4.7 FINDINGS

Finding P-1: DOD processes dealing with the acquisition of C4I systems have not been adequately restructured to account for the rapid pace of development in the commercial information technologies on which such systems will inevitably build.

Acquisition reform is a perennial subject of interest for the DOD. The acquisition process in its traditional forms often takes too long and delivers the wrong products. DOD has undertaken many attempts at making the process more responsive. But the behavior of program directors and managers has evolved little—nor has that of an oversight process established to ensure that every acquisition of significance satisfies the traditional acquisition regulations.

The rapid advancement of commercial information technologies makes available new capabilities for information processing, storage, and communications on a short time scale. Delay in the acquisition process results in a continually expanding delay factor in bringing the power of commercial technology to bear on military C4I requirements. Indeed, the present acquisition cycle virtually guarantees obsolescence upon fielding of military systems when technologies key to their success improve at the rate of an order of magnitude every 5 years.

For military systems (both weapons and C4I) to fully exploit this power, the acquisition process must be shortened. But because DOD no longer enjoys the leverage it once had regarding the development and application of advanced information technology, military C4I requirements must be met through an increasing reliance on technologies provided by the commercial market.

Rapid change in the technologies underlying C4I systems also creates a need, now not met, for regular reevaluations of the balance in resources allocated to weapon systems and C4I, as well as for mechanisms to insert funding to exploit unexpected technological advances.

A second aspect of the acquisition system is that it is particularly ill-suited to C4I systems. Program management and oversight processes are heavily weighted toward metrics associated with historical acquisition methods and tend toward dominance of cost, schedule, and predefined performance measures. Formulation of requirements and acquisition program management and oversight processes result in long acquisition cycles and a bias toward achieving maximum performance. These metrics and approaches are often not consistent with the timely incorporation of commercial technology into C4I systems.

Thirdly, the current acquisition process is premised on the ability of a service to identify a specific system or program to address specific and articulated military needs. While such a premise may be reasonable for weapons systems, it is inadequate for C4I systems for two reasons. C4I systems and especially infrastructure deployments often have greater value in enhancing overall capability for multiple weapons systems than they do for meeting specific needs (e.g., Link 16 or the Global Positioning System). And, because the power of the underlying technologies increases so rapidly, users are often uncertain about how to use that power to fulfill their C4I requirements. Users more often come to understand their requirements through a process of experimentation with prototypes than by deep intellectual analysis conducted on paper. Such a hands-on process for defining requirements runs very much against the grain of the traditional acquisition system.

Finally, personnel in the acquisition process have not been well trained to manage C4I acquisitions or socialized into an information technology culture. For example, program managers receive education and training oriented primarily toward the acquisition of weapons systems rather than C4I systems. The committee notes that many different approaches could be taken to satisfy the objective of acquisition people knowing more about C4I. For example, a separate programming and acquisition system for C4I, analogous to the system used for equipping the Special Forces Command, could easily be staffed by personnel with specialized knowledge of C4I. Even within the regular acquisition system, individuals with training and background in C4I could be trained to perform acquisition, or acquisition personnel could be trained specifically in C4I. Recommendation P-2 makes a specific proposal to address this issue, but the committee believes it is more important to highlight the problem than to specify a solution in detail.

Finding P-2: In many instances, operational processes do not appear to have been reengineered to take full advantage of the capabilities that C4I technology can provide.

Reengineering of existing business processes to take full advantage of new technologies is quite difficult for both the private sector (witness the difficulties that many Fortune 500 companies have in embracing new business concepts) and in the DOD. Nonetheless, when successful, reengineering provides enormous leverage. The competitive arena for the military is not as well defined as that for private-sector enterprises, but it is reasonable to expect that reengineered, technology-exploiting operational processes should enable major competitive advantage in the military, driving revisions of doctrine, smaller logistical footprints, enhanced agility, and a redefinition of the skill set required in the fighting forces. Major C4I system redesigns, like original systems designs, are likely to provide many opportunities for operational process reengineering.

Note that reengineering of processes is not in the end an issue of using the newest and most powerful information technologies. For example, Wal-Mart achieved its remarkable results from reengineering using relatively old mainframe computers. More important than the technology is the fact that successful reengineering eliminates activities that contribute only minimally to the achievement of the overall goals.

In its site visits and briefings, the committee saw a wide range of organizational responses to C4I technology. In some cases, internal processes were being reengineered, new doctrines and modes of combat operations explored, and potential points of high leverage found. In other cases, C4I technology was being applied to automate existing processes within the context of existing tactics and procedures. Some benefits were apparent from these latter efforts, but experience in the private sector suggests that automation of an existing way of doing business quickly yields diminishing returns and seldom results in large (order-of-magnitude) benefits. It is clear that efforts in the former category are difficult to undertake successfully, but such efforts are necessary if the so-called revolution in military affairs is to be even approximated.

The reengineering of every operational process will not necessarily result in order-of-magnitude improvements in efficiency, though some almost certainly would. In order to select processes that are both likely to show substantial improvement from reengineering and also be of high operational military significance, decision makers must draw on individuals with considerable expertise in two areas: the military operational art of war (i.e., doctrine, strategy, and tactics for employment of forces and weapons) and the capabilities made possible by advanced information technologies and C4I systems. Cultivating such individuals is the subject of Recommendation P-1 below.

Finding P-3: The military services have not accorded to information technology and C4I professionals stature comparable to their increasing importance for battlefield operations.

It is widely recognized that the talents and abilities of well-trained and committed technicians are essential to ensure the successful operation of modern military weapons such as jet fighters, warships, and sophisticated ground-based weapons. It is equally vital that the DOD build a suitably sized force of people who have the requisite education, understanding, and skills to translate functional requirements for information technology applications into system solutions, derived principally from the commercial markets where such technology is developed. At this juncture, the DOD is not succeeding in creating either the environment or the incentives to attract and retain such human resources.

Deficiencies in DOD's posture toward C4I professionals occur at two levels. The first is that DOD has not yet found a way to integrate its C4I personnel into combat line elements and to make them fully conversant with military doctrine, strategy, and tactics. Rather, they are regarded as implementers of high-level strategy decisions that are made without their input, and the status and prestige of C4I specialists are not comparable to those of individuals in traditional combat arms specialties. The role of "implementer" was once played by chief information officers of major corporations, but today chief information officers are regarded as part of the senior management and strategy teams in successful corporations. So, too, must the military find a way to integrate C4I personnel into the military establishment, and commanders, planners, and senior leaders must become fully conversant with these forces and the capabilities they bring to the combat domain. In some instances commands do recognize the importance of involving their senior C4I personnel as integral contributors to the decision-making process, but for the most part, the treatment of the C4I personnel in DOD relegates these valuable resources to the second-class status of support, rather than line functions.

The second deficiency is that even for C4I personnel as "mere" implementers, the DOD culture tends to discourage attracting and retaining the necessary engineering, system integration, and applications talent for implementing and sustaining high-technology C4I systems. Information technology talent is a scarce commodity, and as noted in section 4.4, the DOD must compete with higher compensation, advancement opportunity, and job satisfaction in the civil sector for such talent.

Finding P-4: The DOD process for coupling end-user operational needs to C4I systems is inadequate.

The general principle that operational needs should drive the acquisition system is well established within DOD. Under current practices (i.e., the traditional acquisition system), warfighter input (based on the perspectives of the CINCs) is codified in terms of validated military requirements, which are vetted through the Joint Requirements Oversight Council as the basis for new program starts. The acquisition system takes the military requirements and then—some years later—provides for fielding a system intended to meet those requirements. DOD also understands the importance of providing a vision that explicates its long-term goals—

the production and dissemination of Joint Vision 2010 has been a particularly important step forward in this regard.

On the other hand, the operational concerns that initiate the acquisition of a given system often turn out to be just one factor affecting the way in which a system is developed and procured; in practice, repeated engagements between end user and the acquisition system are not often in the critical path from design to procurement. Furthermore, input from the end users—the field commanders—is particularly important in the design and development of C4I systems, because it is more difficult to specify requirements for C4I systems in a form that they can be handed "over the transom" than it is for most weapons systems. In practice, the loose coupling between the acquisition process and warfighter input has a number of weaknesses.

- *It is too slow.* During the time between the articulation of a requirement and the time a system is delivered to respond to that requirement, the actual need may have changed, thus obviating (or more likely, changing) the nature of that requirement. Furthermore, the underlying information technologies almost certainly have changed, perhaps by an order of magnitude in performance, during this time. Thus, in practice, users may well be essentially disconnected from developers.[31]
- *It requires user prescience.* Research in human factors and user interfaces documents the fact that people are often quite poor at specifying in advance the functionality of a computer system that would be most helpful to them, but that "they know it when they see it." The development of a C4I system that involves a human user should call for continuous input into the development process. In addition, because deployment and fielding a system to large numbers of users will inevitably broaden the base from which operational user feedback can be received, the line between "development" and "deployment" should not necessarily be as well defined as implied by the traditional acquisition and fielding model.
- *It is non-adaptive.* Because the primary user input is received only at the start of the process (in the formulation of the military requirement), users must attempt to anticipate all possibilities and scenarios for use without having a good idea of the kind of system that would be truly

---

[31]This is not to say that the acquisition system is always too slow. For example, over the last few years, the Space and Naval Systems Command field activities have developed methods of making extensive upgrades to carriers and carrier battle groups before a deployment. Working at levels below the thresholds of the acquisition system, the Space and Naval Systems Command installs system changes in direct response to the battle group commander's requirements. The entire process operates within the 18-month workup for a battle group deployment. Nevertheless, the fact remains that such speed is the exception.

useful. On the other hand, one of the major advantages of information technology is that it is an enabler for new ways of doing business and approaching problems, new ways that cannot be anticipated without information technology-based systems in hand with which to experiment.

• *It focuses on specific requirements.* An acquisition system that focuses on the satisfaction of specific requirements may well give inadequate attention (in terms of budgeting and management attention) to infrastructure programs that might benefit large numbers of users across different theaters serving different functions.

• *It loses detail.* In the present acquisition system, the articulation of military requirements is a responsibility of the CINCs. Thus, it is only natural that the concerns and frustrations of relatively senior officers are expressed. For example, top-level commanders expressed to the committee during site visits their considerable frustrations with current C4I systems that included slow communications speeds, lack of interoperability with adjacent systems, inconsistent results from different systems, and incompatibility with systems of other forces, domestic and foreign. Theaters such as Korea exhibited difficulties in integrating allied systems with U.S. systems because discrepancies in the budgets for both sides resulted in large gaps in technology deployment. Observed incompatibilities included language problems, data transmission media, and the technical sophistication of information analysis.

By contrast, lower-level personnel have control of a smaller span of C4I systems and are more reliant on a few of them. For example, during Joint Warrior Interoperability Demonstration (JWID) 97, the committee spoke to many warfighters who expressed a need to filter and fuse data from multiple sources and who could greatly benefit from better decision support systems for tactical operations. This observation was repeated in one of the battle centers in Korea, as well as in discussions with field personnel. While such concerns are reflected in some form higher up the chain of command, critical nuances and details available from lower-level personnel are often lost in the abstraction process. While the abstraction process itself is not an unreasonable one (generals DO have a responsibility to filter and abstract the most important pieces of information received from privates and sergeants), the committee was not able to identify a point at which the concerns of lower-level personnel can be fed directly into the development process.

Warfighter input (especially that from a joint perspective) can be diluted when individual services are responsible for the articulation of system performance requirements and specifications. The reason is that while the initial specification of requirements may indeed be joint and operationally based, all development projects entail further refinement of specifications as they proceed (this is especially true if a spiral development process is used). A service perspective—rather than a joint one—is thus automatically present as such refinement proceeds. For C4I systems that are primarily of interest to one service, such a perspective will probably enhance the outcome. But if the system is primarily of interest to a joint commander, or if the system is likely to depend on data provided by C4I systems in other services, a service perspective may well detract from (joint) interoperability and/or full functionality.

This is not to say that service-led C4I programs cannot be successful in producing highly interoperable C4I systems. But the committee believes that such interoperability successes happen because they are managed by particularly dedicated individuals with broad (joint) perspectives themselves, rather than because the acquisition process is optimized to support such outcomes.

Finding P-5: Achieving C4I interoperability is more a matter of organizational commitment and management (including allocation of resources, attention to detail, and continuing diligence) than one of technology.

It is often alleged that procurement of C4I systems that are interoperable with one another would require additional funding. This allegation is undoubtedly true when stovepiped systems are made to interoperate in the later stages of the design cycle. It is also true that designing all systems for interoperability when only some need to interoperate is needlessly expensive. And finally, satisfying a broader set of requirements is often more expensive, all else being equal, than satisfying a narrower set.

On the other hand, designing for interoperability from the start is often less expensive. The reason is that interoperability is a property facilitated by use of common and existing architectures, standards, data definitions, interfaces, and even code. With object-oriented technology, among others, interface requirements can be implemented as class libraries and shared to reduce development time and cost. By reusing existing work (whether manifested as preexisting military technology or COTS technology), major cost savings are possible in the development of a system. Additional savings may be possible to the extent it is possible to off-load the costs of integrating subsystems onto vendors (who are providing COTS products). Most importantly, total life-cycle costs may well be less if the need to hedge against unanticipated needs for interoperability can be reduced, because retrofitting systems for interoperability results in working such problems case by case, providing expensive curative rather than inexpensive preventive medicine.

In addition, experience with corporate mergers in the commercial world suggests that consolidation of information technology infrastructures can result in considerable cost savings, even when the prior infrastructures were incompatible. Additional costs result from the need to make two infrastructures interoperate, but in the long run, more money is saved because excess capability in the infrastructures can be used more efficiently. To illustrate this principle in a military context, commanders are often faced with the paradox of wanting more bandwidth even as existing channels are not used to capacity, as the committee saw in Korea recently (in 1997) and as was especially true in the Gulf War in 1991. The reason for this paradox is that existing channels are stovepiped for the exclusive use of one system or another, thus rendering their excess capacity useless to other parties who could make use of it.

Nor does achieving interoperability require the development of new technologies. Interoperability problems result from human decisions to design systems with different specifications. Technology can sometimes be useful in helping to reconcile differing lower-level specifications (e.g., different frequencies or different voltage levels or different protocols) automatically, but no technology can be expected to automatically reconcile differing human judgments about higher-level issues (e.g., those related to data semantics and information flows, or those relating to judgments about releasability to foreign nationals).

The committee believes that senior DOD leaders, both civilian and military, take interoperability challenges quite seriously. But DOD lacks a process for establishing a culture supportive of C4I interoperability that will outlive today's senior leaders. Absent such a culture, DOD efforts to promote and enforce interoperability will be fragile. For example, consider the fact that the C4I apparatus within DOD is subject to nearly constant change (both threatened and actual), a fact that leads to the conclusion that the C4I constituency is unstable and constantly under fire. A constantly shifting bureaucratic base does not give confidence that high-level management attention to C4I issues can be sustained. If so, individuals associated with the program can simply wait things out until the next bureaucratic rearrangement. A second example is the proliferation of organizations within DOD with some responsibility for interoperability. In the committee's view, the very existence of many such organizations strongly suggests that none of them work very well to achieve their interoperability goals.

Given the unavoidable fact that the senior DOD leadership turns over on a time scale short compared to the time that it takes for major cultural change to occur, DOD must rely on the creation of an enduring process to promote its C4I goals, especially interoperability, rather than on the services of any particular set of individuals. Moreover, because oversight is inherently time-consuming (because of the assumption that things may be going on that may not be fully consistent with organizational goals), this process must be based on the establishment of cultures and incentives that support interoperability, rather than oversight alone.[32]

4.8 RECOMMENDATIONS

Because the C4I systems in question are constantly in flux, understanding how to manage technological change assumes at least as much importance as the technologies or the architectures themselves. The effectiveness and efficiency with which joint military operations can be conducted will depend heavily on how well the services can collaborate and sustain that collaboration over time before the battle. Such peace-time collaboration, among fiercely independent groups like the services, however cannot be dictated, legislated, or simply announced. It will require the establishment, within the services and the DOD, of a supportive environment that can foster continual, effective, efficient independent collaboration and the development and use of internal systems that can support such collaboration.

DOD must alter its military and civilian culture in ways that are commensurate with the importance of C4I to its future vision. Organizations often say that their most important asset for the future is their people. The reason is that it is people who implement policies and carry out the day-to-day operations of the organization. Without good people, the best plans cannot be executed effectively. But organizations whose prevailing culture encourages behavior that does not support management goals also find that the plans of management are not well executed.

The DOD is a large organization, and many aspects of its culture could be changed. As before, the committee focuses here on several areas that it believes provide high leverage. In addition, DOD must change certain key aspects of the acquisition system for C4I systems if the full potential of new C4I systems and technology is to be exploited.

As in previous chapters, the recommendations in this chapter on DOD process and culture are cast in terms of *what* the committee believes should be done, rather than specifying an action office. The argumentation for each recommendation contains, where appropriate, a paragraph regarding a *possible* action office or offices for that recommendation, represent-

---

[32]The longevity of the process is particularly relevant to the completion of the various joint architectures. Constructing these objects for C4I is something that will span the lifetimes of more than one tour of duty. Success seems critically dependent on devising a process that ensures their construction, and the committee did not observe any evidence that this process was in place, or that people recognized that such a process was needed.

ing the committee's best judgment in that area. However, this action office (or offices) should be regarded as provisional, and DOD may well decide that a different action office is more appropriate given its organizational structure.

Recommendation P-1: The Secretary of Defense, working with the service Secretaries and the Chairman of the Joint Chiefs of Staff, should establish in each of the services a specialization in combat information operations, provide better professional career paths for C4I specialists, and emphasize the importance of information technology in the professional military education of DOD leadership.

Today, the treatment of the technical force in DOD relegates these valuable resources to the second-class status of support, rather than line functions. If it is true that information is critical to the prosecution of modern warfare, and that information dominance can provide the operational military advantages of large forces without incurring their costs, then specialists in C4I systems must be better aligned with those in the mainstream operational community.

Better alignment begins with unified, joint and component commanders who have a good understanding of how best to exploit information technology and C4I to enhance military operations (e.g., rapid change and new capabilities thereby enabled). Developing dual competencies—both technological and operational—among military leaders is likely to require changes in their professional military education throughout an individual's career (perhaps including rotational operational tours for "information systems" personnel and information systems tours for operational personnel). Such changes would focus greater attention on the role and potential impact of C4I and information systems on the operational art of war. Information system employment must become a first line combat function, just as is employment of combat forces and weapons.

This means that the C4I specialists must not be regarded as "geeks off to the side" or as mere implementers but as individuals who are trained in and involved with doctrine, training, and operations—full members of the combat operations team that are fully conversant with the operational employment of military forces. C4I specialists should not only be knowledgeable about the relevant technical disciplines, including communication systems operations, information warfare, information security, and so on, but also be engaged as military operators, involved in combat operations just as completely and widely as today's combat infantry, armored, sea- or air-power operators. It is, therefore, essential that the C4I specialists be trained in the doctrine, strategy, tactics, and combat use of military forces, and be fully integrated into the combat units and operational planning elements of the military forces.[33] Professional military education opportunities provide many forums in which traditional combat operators and C4I specialists can learn about the specializations of the other. (The Army Signal Corps is an example of making the technology people part of the operational team.)

Other steps that can be taken to support this recommendation include:

- *Increased promotion opportunities and recognition.* C4I specialists, as well as combat arms specialists, should have the opportunity to compete for and attain command of combat forces and advancement to the most senior positions of responsibility. Note that the flip side of greater promotion opportunities and status is increased responsibility: system administrators need to have more of an "operator on the front lines" mentality that exposes them to an environment different from that of administrators working for a bank, and the C4I specialists in all decision-making roles must be held accountable for operational outcomes just as all other team members are held responsible. Furthermore, DOD should develop ways to celebrate the excellence and importance of the C4I specialists publicly and graphically even apart from expanding their upward mobility.

- *Higher pay scales ("proficiency pay" or "incentive pay" or "special pay") for C4I specialists.* While it is true that military service is a privilege, the fact remains that for disciplines in which the private sector competes with the military for talent, the higher rates of compensation found in the private sector are a powerful draw for many of those with talent. Additional compensation for C4I specialists that partially makes up for the private-military pay gap would help to reduce the outflow of talent from the military, especially at the lower levels.[34]

---

[33]In this regard, the combat information specialists of the future are likely to share certain characteristics of good intelligence analysts today. In particular, the value of intelligence advisors to a commander is significantly enhanced when the intelligence analysts understand the significance of information about the enemy in the context of the commander's intent and how the commander is likely to want to use that information. And, he advises the commander on what can and cannot be done given his knowledge of both "red" and "blue." Such ability depends on being quite knowledgeable about the doctrine and capabilities of friendly forces as well as about those of the enemy. Combat information specialists will also have to be able to collect and integrate information from many sources, just as intelligence analysts do today.

[34]Additional compensation is made available to about 43% of military personnel who receive special and incentive pays offered as inducements to undertake or continue service in a particular specialty or type of duty assignment. Examples of these pays include Jump Pay, Sea Pay, Submarine Duty Pay, Flight Pay, Imminent Danger Pay, medical and dental officer pays, and various enlistment and reenlistment bonuses (see Office of the Under Secretary of Defense for Personnel and Readiness, available online at <http://dticaw.dtic.mil/prhome/paybenef.html>). The pays for medical and dental officers are particularly relevant, because they are offered primarily to reduce the salary differential between military service and the private sector.

The military services are the appropriate action office for establishing a specialization in combat information operations. The service training and doctrine commands, as well as the various schools that provide professional military education, are the organizations through which combat operators can become more familiar with C4I and information technology. The Joint Staff, specifically the Directorate for C4 Systems and the Directorate for Operational Plans and Interoperability, would be an appropriate office to conduct a review of this area and formulate recommendations for improvements where needed. And, the criteria that service promotion boards examine in determining promotions will be critical in promoting a more C4I-knowledgable military. Expansion of the possible career paths for C4I specialists is a function of the Under Secretary of Defense for Personnel and Readiness.

Recommendation P-2: The Under Secretary of Defense for Acquisition and Technology should train its civilian and military personnel who participate in the acquisition of C4I systems to understand the difference between C4I systems and weapons systems.

Program managers must understand the intrinsic differences between C4I and weapons technologies, and they must be able to argue the significance of those differences in front of acquisition boards and oversight councils that are more accustomed to dealing with weapons systems. Today, conservative "by the book" approaches that are better suited to long-lived weapons systems are regularly applied to C4I systems, even though the existing acquisition process allows considerable flexibility in the management of a C4I program.

If program managers are to advocate non-traditional approaches to acquiring a C4I system, testers and evaluators must understand the impact of these non-traditional approaches and refrain from judging C4I systems according to traditional criteria. For example, they must understand from a testing perspective the ramifications of evolutionary acquisition.

One appropriate forum for such education would be the Defense Systems Management College of the Defense Acquisition University. The Defense Systems Management College provides systems acquisition education and training for the people responsible for acquiring weapon systems. As such, it offers courses of study that are designed to prepare selected military officers and civilians for responsible positions in program management and other associated acquisition functions.

Recommendation P-3: In order to explore and develop ("incubate") new ideas for the use of information technology to support military needs, the Secretary of Defense should establish an Institute for Military In-

formation Technology either as a free-standing unit or by expanding the charter of an existing institution.

Experience from the commercial sector demonstrates that creative information technologists flourish in an entrepreneurial culture that encourages and rewards intellectual risk taking. To a large degree, this phenomenon results from the fact that risk takers are allowed to keep the fruits of managing risk successfully. But since the DOD culture does not generally allow the risk takers to reap such benefits, shaping such a culture within the military requires that its leaders help absorb those risks for those with good ideas, regardless of their level in the hierarchy. Indeed, the committee believes that all levels of the DOD/service hierarchy contain individuals with good insights about existing problems, ideas about how to fix those problems, and innovative concepts about how C4I technology could be used to improve military effectiveness. But because of the traditional military command structure, those at lower levels of the hierarchy face considerable risk if they challenge the conventional wisdom.

One example of where risk absorption is essential is in the receiving of operational feedback from end users. The end user must feel comfortable about being honest in the evaluation (without negative feedback from the designer), while the designer must be protected from the penalties of negative feedback, which should be used to direct system improvements.

Risk absorption should not be confused with a lack of accountability. Risk absorption deals with different kinds of risks: the risks of proposing new ideas, the risks of undertaking ventures or experiments that may fail. In the first case, proposal and advocacy of a new idea do not raise questions of accountability. In the second case, the individual or organization in question should be assessed on the basis of whether or not the experiment was well founded in light of what was known at the time the decision to proceed was made.

A major purpose of the proposed institute is to facilitate creative intellectual risk taking. Toward this end, it would bring together for extended periods of time combat operators, military information technologists, and civilian information technology experts from academia and industry in an environment where innovative ideas for using information technology to support military needs could be explored relatively freely and with minimal personal risk. Innovation would be encouraged by an institutional culture that applauds success and provides for soft failure.

A key element of the proposed institute is the synergy between technologists and the military operators. The technologists provide the supply side—what can be done with information technology. Such information, especially if it is visionary, can influence markedly the commander's view of what will be possible in military operations. The operators provide the demand side—the commander's "druthers," i.e., what he would like to be able to do militarily—information that can stimulate the development of new applications and perhaps new technologies. Technologists will learn from the operators in side-by-side contact and in understanding lessons learned from demonstrations, experiments, exercises, and operational deployments.

The educational dimension of the institute would be approximately that of advanced graduate education in the private sector—learning through problem-based work rather than courses (as is more typical of undergraduate education). Thus, its educational intent would be to share knowledge rapidly and adapt what it teaches to the changing world in a timely manner. In this fashion, it would not operate as a training command, in which courses focus on established doctrine (which—quite properly—takes a long time to evolve). An educational dimension structured along such lines would also enable the institute to provide ongoing support for a career path for C4I specialists.

The institute would also serve as a "think tank" responsive to the services (especially the doctrine commands) and to the Joint Chiefs of Staff. The output of the institute would be both reports and "prototype" or "proof-of-concept" demonstrations. (In this latter output, the institute would differ from traditional think tanks.) As a rule, the institute would not develop technology on its own, instead focusing on the potential adaptation and use of commercial off-the-shelf capabilities in military information technology applications. The technology work undertaken by the institute would thus focus primarily on integration and "stitching together" COTS components to serve military needs.

It is expected that the institute would connect closely with a variety of different institutions and activities:

• Training and doctrine commands and the Joint Battle Center, through which the institute could facilitate a close coupling between service-based strategy and analysis and joint C4I experimentation;
• Service and defense agency research and development efforts in information technology, and the service development laboratories, through which the institute could keep abreast of current C4I developments;
• The Joint Warrior Interoperability Demonstrations, through which the institute could demonstrate in-house work of its own and/or facilitate appropriate work originating in other DOD or contractor bodies;
• The Joint Staff (especially the Directorate for C4 Systems and the Directorate for Operations), through which the institute could couple to operational concerns; and
• The various war colleges, through which the institute could help to

develop the intellectual basis for a broad educational program on C4I issues, particularly for military leaders.

With a stated mission to ensure excellence in professional military education and research in the essential elements of national security, the National Defense University is one possible location for the proposed institute, though it would have to extend itself to engage technologists and system developers. The Joint C4ISR Battle Center is a second possible location, though it would have to extend itself to embrace a research and education function that it currently does not have.

Recommendation P-4: The Assistant Secretary of Defense for C3I and the Under Secretary of Defense for Acquisition and Technology, working with the service Secretaries and the Chairman of the Joint Chiefs of Staff, should direct that as a general rule, every individual C4I acquisition should (a) use evolutionary acquisition; (b) articulate requirements as functional statements rather than technical specifications; and (c) develop operational requirements through a process that includes input from all the services and the CINCs.

Over the time scale of a typical military C4I program, the applicable technology underlying the program, as well as operational requirements for its use, the doctrine that governs its operation, and the world and local environments in which it must operate, can be expected to change dramatically. Large increases in performance mean that features or capabilities desired by users that may have been unrealistic at the start of the program (i.e., when the requirements are first defined) may become more realistic later in the program. Moreover, the nature of the relationship between users and C4I systems is such that users are often unable to foresee how a system might be used without actual operating experience. However, once given that operating experience (something that requires a functioning system), they are in a much better position to articulate other needs and requirements that they did not realize they had. Waiting for a 100% complete statement of requirements that the system will eventually have to meet is a recipe for radically increasing costs and extensively delaying system deployment.

For these reasons, an "80% solution"—an evolutionary acquisition—to the functional requirement, followed by effective preplanned product improvements is not unreasonable as the initial statement of requirements. Such a formulation would encourage commercial technology application and dramatically reduce the cycle time for developing new C4I systems.

An important corollary is that in many cases it is necessary for program plans to state only functional requirements. Indeed, overspecification of the design limits the ability of a supplier to find better or more cost-effective ways of implementing the system. The major exception to this general principle is in the specification of interfaces to other systems. Because these interfaces are essential to achieving interoperability, a high degree of detail is appropriate in specifying them. Such detail should be derived from the operational, technical, and systems architectures that describe the system in question and how it relates to other C4I systems.

Finally, if the intent of the Chairman of the Joint Chiefs of Staff[35]—that all C4I systems developed for use by or in support of U.S. forces are by definition to be considered for use in joint operations—is to be met, the requirements definition process should be under the control of a group that represents the interests of all stakeholders. As a general rule today, requirements are initially specified by the service programmatically responsible for a system to be acquired; other stakeholders such as the CINCs or the Joint Chiefs of Staff have opportunities for input, but primarily in later stages of program review when the system has been largely defined. Furthermore, while the requirements for some programs are vetted through the Joint Requirements Oversight Council, the Defense Acquisition Board, or the Major Automated Information Systems Review Council, these bodies deal only with programs exceeding some (relatively large) dollar threshold (and the magnitude of a C4I program is not a good indicator of its operational importance). And, the fact that these bodies perform a review and oversight function for many programs means that they are limited in the attention that they can give to any specific program. The committee believes that a process that ensures input from the CINCs and inter-service input in the initial formulation, as well as the review of requirements, increases the likelihood that the requirements that a system is designed to meet will in fact satisfy needs for interoperability and jointness.[36]

Note: This recommendation does not call for the establishment of joint offices for program management. While under some circumstances a joint program office for a C4I program may be appropriate, a joint program is dependent on the services for the monetary support, staffing, contracting,

---

[35]Chairman of the Joint Chiefs of Staff Instruction 6212.01A

[36]It can be argued that the involvement of all services and the CINCs in the formulation of C4I requirements will simply result in an ever-expanding list of requirements that would lead to higher unit costs. For example, if an advocate of certain requirements has no responsibility for supporting a system to meet them, the "wish list" becomes a free good that is easy to abuse. It therefore falls to program managers to discipline the process of formulating requirements so that the list does not continue to expand. (One approach might be to require programmatic contributions from other services to fulfill requirements that are associated with the needs of those other services.) In any event, a broad perspective on C4I requirements is intended by this recommendation.

and purchasing authority needed to execute an acquisition. Such dependency frequently leads to multiple inadequacies in program execution and can make the joint program less effective due to the inability of a joint program director to control service support for his program.

Because the Under Secretary of Defense for Acquisition and Technology and the Assistant Secretary of Defense for C3I have the ultimate responsibility for acquisition matters related to C4I, those offices are the appropriate ones to take action. The policy promulgated must require explicit justifications for approaches to acquisition that do not call for evolutionary acquisition, must be observed by all service acquisition arms, and must specify that all requirements contained in program documents for all C4I programs and all C4I within weapons systems be stated as functional statements.

Recommendation P-5: The Secretary of Defense should seek, and the Congress should support, an appropriate level of budgetary flexibility to exploit unanticipated advances in C4I technology that have a high payoff potential.

As new commercial information technologies and applications emerge that can significantly improve military capabilities, management and budgeting approaches must be flexible and responsive if timely acquisition of fast-paced information technological developments is to succeed. High-value C4I applications that emerge from an advanced concept technology demonstration (ACTD) or a demonstration such as those in the Joint Warrior Interoperability Demonstrations are all too often "orphaned" in relation to the regular acquisition track, and follow-through has been difficult in the past.

The reason is that the normal planning and budget process programs funds years in advance. Thus, some "offline" funding mechanism is required to cover unanticipated needs.[37] Furthermore, even if an ACTD does not enter the mainstream acquisition process, funding streams are

---

[37]Today, mechanisms available to cover unanticipated needs include reprogramming authority (which, up to a certain limit, can be exercised without congressional approval) and emergency or supplemental appropriations (which require congressional action). By definition, reprogramming funds one program at the expense of another, and so can be expected to generate considerable controversy. Furthermore, from the standpoint of the program being used as the funding source, reprogramming adds considerably to the difficulty of managing it. Supplemental appropriations leave previously authorized/appropriated funding streams intact, but take time to happen and are procedurally cumbersome. The committee does understand congressional concerns about exercising oversight responsibilities, but the legislative time scale is long compared to the time scales that characterize the emergence of new opportunities in information technology.

needed to ensure that leave-behinds from ACTDs are compatible with the other systems where they are deployed, and are maintainable and supportable.

Of course, a C4I ACTD that is developed independently of various requirements to support interoperability and security is unlikely to be adequately interoperable or secure. Thus, ACTDs should be developed in conformance with such requirements, even if such development increases the initial research and development cost. C4I ACTDs that are not adequately interoperable or secure are not likely to have significant "leave-behind" operational utility in the long run in any event, so that funding streams for such ACTDs are not needed.

Given the tension between effective budgetary oversight and budget flexibility, the senior leadership of the DOD must take the lead in expressing the need. While budget flexibility is always regarded as desirable by those whose budgets are being overseen, the time scales on which useful applications of C4I can emerge is much smaller than the characteristic time scales of the DOD budget, making such flexibility particularly important in the C4I domain. For example, one approach for increasing flexibility that might warrant consideration (though the committee is not specifically recommending it) is to increase for C4I programs (and for C4I programs only) the current thresholds for budget reprogramming below which the DOD can take action without explicit legislative approval.

Recommendation P-6: DOD should put into place the foundation for a regular rebalancing of its resource allocations for C4I.

C4I is a fundamental technological underpinning of information superiority. If DOD is serious about its commitment to U.S. information superiority on the battlefield of the future, it must be engaged in a thoughtful and continuing examination of the resources it allocates to C4I. The outcome of such examination may support the beliefs of different constituencies within DOD about the proper future trajectory of C4I resources. Some believe that the fraction of the DOD budget devoted to C4I should increase significantly in the future; others believe that the amounts should decrease, and still others say it should remain about the same. The committee is explicitly silent on whether the budget is appropriately balanced today among readiness, weapons, force structure, and other types of military spending, but it does note that an increase in the fraction of the budget devoted to C4I necessarily entails trade-offs against these categories.

The committee believes that DOD would increase the likelihood of making sensible budget decisions about C4I if it put into place the foundation necessary for undertaking a rebalancing of C4I vis-à-vis weapons

PROCESS AND CULTURE   239

and force structure as part of the regular budget process. Key elements of this foundation are the focus of the following two sub-recommendations.

Recommendation P-6.1: The Under Secretary of Defense (Comptroller) should explicitly account for C4I spending as a whole in DOD's budget process.

As noted in Chapter 1, C4I is not a budget category within the annual DOD budgeting process. In the absence of such information, it is left to a large extent to the services to determine their own C4I priorities and how those weigh against their needs for force structure and weapons procurement. Input from the Joint Chiefs of Staff provides an opportunity to take a more integrated perspective, but without knowing what is being spent by all of the services on C4I in any given year, it is obviously difficult to take a defense-wide perspective on the level of overall spending.

It is true that the most recent Quadrennial Defense Review (1997) performed a cross-walk through the budget to determine how much was being spent on C4I. However, 4 years is far too long a time to elapse between the points at which the overall spending on C4I is understood. While it does not make sense to build a new C4I plan every year, plans must be updated on a time scale comparable to that for significant progress and change in the underlying technologies. This time scale is much closer to 1 year than 4 years.

Whether an overall assessment of spending on C4I should include C4I that is embedded into weapons systems is an open question. On the one hand, weapons systems and command decisions will rely on certain *capabilities*, whether they are provided by systems that are programmatically designated as C4I systems or not. Thus, from an analytical standpoint, the programmatic category should not matter. On the other hand, extracting the costs of embedded C4I from the overall costs of a weapons system in which it resides may be quite difficult and prone to error. In particular, data not subject to a consistent reporting scheme across all weapons systems programs may cause problems for one program vis-à-vis another. Furthermore, weapons systems program managers may well be reluctant to explicitly call out the cost of C4I for fear of increasing its visibility to budget auditors.

Whatever definition of "C4I" is adopted, it must be governed by consistent accounting rules. These rules would address questions such as whether or not to include sensors physically carried by a platform (e.g., the radar built into the F-22), sensors operating in close proximity to a weapon (e.g., the radar associated with the Patriot missile system), and off-board sensors used to support precision strike operations (e.g., sensors carried on platforms such as a JSTARS aircraft).

Because the Under Secretary of Defense (Comptroller) is responsible for supervising and directing the formulation and presentation of defense budgets and establishing and supervising the execution of uniform DOD policies, principles, and procedures for budget formulation, it is this office that must take the ultimate responsibility for a more frequent accounting of C4I expenditures and for using this accounting in establishing spending priorities. Of course, it is expected that the comptroller would work closely with the Assistant Secretary of Defense for C3I in conducting such an accounting.

Recommendation P-6.2: The Joint Chiefs of Staff should develop and use measures of military effectiveness that can be used to assess the contribution of C4I to military effectiveness.

An increase in the fraction of the budget devoted to C4I necessarily entails trade-offs against other modernization, readiness, and force structure. Given that these costs will likely have major implications for force effectiveness, DOD should be confident that the benefits from more C4I resources are strong enough to provide a net positive result if it decides to move in that direction. Quantitative measures of military effectiveness will thus be necessary to support a continuing process of rebalancing investment among C4I, weapons, and force structure (and among C4I systems themselves). Furthermore, quantitative measures can also help to inform the judgment of senior military leaders about how the capabilities offered by C4I can best be exploited in conducting military operations (i.e., in the formulation of military doctrine).[38]

Some indications of the contribution that C4I can make to military effectiveness are known from simulation and modeling as well as experiments. However, authoritative, accepted models typically do a poor job of representing C4I capabilities and performance in a realistic way, and C4I-oriented models that at least partially compensate for this shortcoming are generally neither comprehensive nor broadly accepted. Most commercial communications systems and process control systems do use mathematical models and simulations in some fashion. Sometimes relatively simple models and measurements result in substantive improvements. The same should apply to C4I systems.

---

[38]This argument is not to say that all aspects of warfare can be quantified with precision. In particular, quantitative measures of military effectiveness that support force structure and investment decisions are very different from statistics that measure operational battlefield encounters. Emphasis on the latter leads to a "body count" mind-set that may have minimal relevance to actual military outcomes, and to managers "making their numbers" without regard for the overall objective.

To support intelligent decisions about investment and doctrine, tools are required at several levels:

- Measures that characterize the performance of C4I systems such as decreased latency of situational information at all echelons;
- Measures that characterize the contribution of C4I systems to particular military operations such as improved rates of fire, more effective expenditure of firepower, or increased ability to place targets at risk; and
- Results from force-on-force simulation and exercises, which enable assessment of overall contributions afforded by C4I as well as new doctrine that exploits C4I capabilities.

Analysts have sought for many years to develop measures of military effectiveness for C4I, and the committee recognizes the difficulty in developing them. But the difficulty in developing such measures should not be used as an excuse for ignoring them. Measures of military effectiveness for C4I, including intermediate measures for interoperability and security, can be defined, however incomplete and overly simplistic they may seem initially, and systematically used to measure progress in achieving the DOD's objectives for C4I. In some cases (perhaps such as interoperability—see Chapter 2), scorecards will have to suffice initially.

Finally, measures of military effectiveness and simulations and exercises must be based on scenarios with operational significance. They must be based on real military requirements and independently developed rather than developed specifically to showcase particular C4I systems or concepts.

The Joint Staff Directorate for Operations is the most plausible office to take action to support this recommendation because it has the closest connection to operational scenarios and deployments. Because a considerable amount of research and development in this area may be necessary (indeed, new theories of warfare may be needed), the Directorate for Operations may well contract significant work with various analytic organizations (e.g., RAND, the Institute for Defense Analyses, MITRE).

Recommendation P-7: The Secretary of Defense, the Chairman of the Joints Chiefs of Staff, the CINCs, and the service Secretaries should sustain and expand their efforts to carry out experimentation to discover new concepts for conducting information-enabled military operations.

Experimentation within the DOD context is analogous to business process reengineering in the private sector. Both seek radically new ways of doing things that create value and advance the ability of the organization to conduct military operations or to make money. Experimentation and business process reengineering can take place at many different scales—from how a combat operations center does its work to how Army corps and Air Force wings and Navy battle groups fight battles. Some may be relatively costly (e.g., the Army's Advanced Warfighting Experiment), others less so (e.g., the Air Force's Expeditionary Force Experiment).

Sometimes small-scale experiments that are less inexpensive lay the groundwork for success in larger experiments. For example, it is appropriate for the Army to have conducted small-scale experiments with digitizing battalion-sized forces before similarly equipping a full brigade. Even larger-scale experiments may be cost-effective in the long run if they help make the right investments and avoid the wrong ones. However, it is also important to note that the reengineering of business processes can have a high impact even with relatively low expenditures on technology (as the Wal-Mart experience demonstrates).

A number of techniques have been used to facilitate process reengineering. For example, the use of integrated process teams in key functional areas could be used to develop reengineered processes to go along with the use of new (or existing) C4I systems. Process "tiger teams" can be used in the field to "walk the process" and talk to individuals involved in a process at every level; such teams can be useful not only in discovering reengineering opportunities, but also in gaining understanding and support from the community that is the object of reengineering.

Reengineering often entails disincentives. Specifically, reengineering of business processes often results in many fewer people being needed to accomplish the same end result. The people who might be displaced by reengineering have vested interests in resisting it (and there is also the non-trivial emotional factor of being deemed "irrelevant"). Furthermore, the larger organization of which these people are a part may not wish to give higher authority a rationale for reducing its personnel levels (or budget). The fear is that if an organization saves money through reengineering, its budget will be cut, the savings directed elsewhere, and the organization left vulnerable to the risks of innovation. Under such circumstances, assurances that the organization will not face such losses can play an important role. (In the DOD context, such assurances must come both from the senior leadership of the DOD and from the Congress as well.)

Significant efforts to support experimentation are under way today. For example, a major step in this direction has been taken in the designation of the U.S. Atlantic Command as the leader in joint experimentation, with a new organization in the Joint Chiefs of Staff for experimentation consisting of approximately 400 staff. The Army's Advanced Warfighting Experiment has been strongly supported by the DOD and the Congress.

In a recent initiative, the U.S. Pacific Command recently conducted an experiment to assess the value of the Virtual Information Center to support the needs of the theater commander-in-chief and joint task force commander in humanitarian assistance and disaster relief operations. The Joint Battle Center, a creation of the Joint Chiefs of Staff, provides the combatant commands, at the joint task force level, with a joint capability and experimental environment that will be a forcing function for joint C4ISR capability and will foster rapid, near-term insertion of C4ISR technology. The Joint Battle Center will be a learning and experimentation center for the warfighter and the technologist, supporting Joint Vision 2010 and the requirements of CINCs for C4I capability.

Nevertheless, it is all too easy to fall back to "business as usual" when faced with budget pressures. Experimentation is undeniably expensive, and failure is to be expected from time to time. Well-meaning critics who focus on the cost and possible failure of particular individual experiments may wind up doing more damage than good in the long run. Fortunately, such criticism is rare today, but the committee lays down a marker for the future.

The organizations that support experimentation need no exhortation that experimentation is a good thing to do. But in the face of budget pressures to cut back on experimentation, the Secretary of Defense, the Joint Chiefs of Staff, the CINCs, and the service chiefs will have to strongly uphold the value of investing in the future.

Recommendation P-8: DOD should develop and implement a set of management metrics that are coupled to key elements of C4I system effectiveness.

Achieving large-scale cultural change in an organization requires commensurate change in management metrics. Indeed, a maxim of quality management is "if you can't measure it, you can't improve it." Metrics, a major motivator of human behavior, have been demonstrated to be an essential element of making improvements, and are the base for driving continuous progress.

In general, management metrics focus on the characteristics or performance of an organization, and are used by senior management to assess the effectiveness of the organization and its leadership. The committee is aware of some areas where DOD is attempting to apply management metrics to drive cultural change within the department.[39]

---

[39] One example would be DOD's formulation of criteria (still in process) for holding unit commanders responsible for information security practices in their commands, as discussed in footnote 27.

A range of such management metrics are required to assess and drive change associated with exploiting the full leverage of C4I in warfighting. Metrics aligned with such key areas as interoperability, security, and overall rate of implementation, as well as such associated elements as training, and skill resource levels, are called for. These metrics must be as quantitative as possible, though in some cases judgment-based ratings will have to be used. The metrics should be applied to units as well as commanders at higher echelons in a manner consistent with their responsibilities. Box 4.3 provides some examples of management metrics for gauging progress toward C4I implementation goals.

Appendixes

Appendix A

List of Site Visits and Briefings

SITE VISITS OF THE COMMITTEE

Joint Warrior Interoperability Demonstration 97, Tidewater, Virginia, July 1997
Ulchi Focus Lens, Korea, August 1997
Force XXI Division Advanced Warfighting Experiment, Fort Hood, Texas, November 1997
Blue Flag 98, Eglin Air Force Base/Hurlburt Field, Florida, February 1998
National Security Agency, Fort Meade, Maryland, May 1998
Electronic Systems Command, Hanscom Air Force Base, Massachusetts, May 1998

BRIEFINGS TO THE COMMITTEE

June 1997

C4I Acquisition and Technology
Noel Longuemare, Acting Under Secretary of Defense for Acquisition and Technology

Fundamentals of Command and Control
Dave Alberts, National Defense University

NATO C3 Issues
Loren Diedrichsen, NATO C3 Agency

DOD C4I Issues and Future Challenges
James Soos, Cheryl Roby, Dennis Nagy, Office of the Assistant Secretary of Defense for C3I

Joint Vision 2010
Colonel Fred Stein and John Garstka, Joint Staff C4 Systems Directorate

The Legislative Framework for C4I
Anthony Valletta, Principal Deputy Assistant Secretary of Defense for C3I

September 1997

Defense Information Infrastructure/Common Operating Environment/Global Command and Control System
Rear Admiral John Gauss and Dr. Frank Perry, Defense Information Systems Agency

Network-Centric Warfare
Vice-Admiral Arthur Cebrowski, U.S. Navy

Joint Warrior Interoperability Demonstration Wrap-up: Lessons Learned
Captain Dennis Murphy, U.S. Navy

Operations Other Than War: Agile Lion
Lieutenant Colonel Chris Weldon, U.S. Marine Corps

Service C4ISR Representatives
—Lieutenant General Bill Campbell, U.S. Army
—Lieutenant General William J. Donahue, U.S. Air Force
—Colonel John Douldry, U.S. Marine Corps

Special Operations Command
Jim Cluck, Special Operations Acquisition

December 4, 1997

The DOD Acquisition Process for C4I
Dr. Margaret Myers, Office of the Assistant Secretary of Defense for C3I, and Mr. Ronald Mutzelburg, Office of the Under Secretary of Defense for Acquisition and Technology

APPENDIX A 249

The Joint Interoperability Test Command
Colonel Tom Andrew, Deputy Commander, and Mr. Butch Caffall, Technical Director

Advanced Concept Technology Demonstrations and the Theater Precision Strike Operations
Advanced Concept Technology Demonstration
Mr. Joseph Eash

Theater Precision Strike Operations
Mr. Bruce Zimmerman, U.S. Army/Office of the Assistant Secretary of the Army for Research, Development, and Acquisition

Joint Continuous Strike Environment
Mr. John Osterholz, Office of the Assistant Secretary of Defense for C3I

Joint Theater Air/Missile Defense
Richard Ritter, Ballistic Missile Defense Office

Eligible Receiver
Captain Jake Schaftner, Joint Staff

March 1998

Roundtable discussion with Lieutenant General Muellner, Lieutenant General Kadish, and Brigadier General Nagy to discuss Air Force C4I acquisition issues

Presentation by Lieutenant General Dennis Buchholz (Director, C4 Systems Directorate)

Appendix B

Summary of Relevant Reports and Documents

B.1 STUDIES OF THE OFFICE OF THE SECRETARY OF DEFENSE AND JOINT CHIEFS OF STAFF

B.1.1 *Report of the C4ISR Integration Task Force*, 1996

Background: The C4ISR Integration Task Force (ITF) issued its report on November 30, 1996.[1] The ITF was created in 1995 by the Deputy Secretary of Defense to "define and develop better means and processes to ensure C4I capabilities most effectively meet the needs of our warfighters." Although other efforts are examining C4ISR integration and interoperability, the ITF was formed to address these issues from a broader perspective. The ITF's goals were to (1) set an aim for the C4ISR functional area by creating a defense-wide C4ISR "Strategic Vision and Guiding Principles," and (2) improve the "processes (architectures, requirements, resource allocation, and acquisition) that impact C4ISR capabilities needed by the warfighters and decision makers."

C4ISR Vision: The Integration Task Force developed a C4ISR vision for the 21st century, based on concepts identified in *Joint Vision 2010* and *C4I for the Warrior:* "Warriors, and those who support them, generate, use,

---

[1]C4ISR Integration Task Force. 1996. *Report of the C4ISR Integration Task Force*, Department of Defense, Washington, D.C.

and share the knowledge necessary to survive and succeed on any mission."

C4ISR Guiding Principles: In order to achieve this vision, C4ISR capabilities generally need to be effective, affordable, and adaptable. In addition, the C4ISR capabilities and processes should be inherently joined and coalition-capable; interoperable; tightly coupled to requirements; secure and available to authorized users; robust and survivable; doctrinally agile; widely available and timely; able to share knowledge that can be tailored to the need; cognizant of the reality of chaos and able to deal with uncertainty; self-aware and self-healing; able to share language; able to keep pace with evolving technology; mobile and continuous; adaptable and adaptive; conformable to standards; easy to use, effective, and fast; innovative; and based on learning, collaboration, and empowerment.

ITF's Recommendations: The summary below focuses on the ITF's 13 major recommendations, which are organized into five categories. These recommendations and associated strategies, as well as action offices, time lines, and targets where appropriate, are discussed in detail of Chapter 5 of the ITF's report.

1. *Manage and guide:* A common strategic direction needs to be established to guide C4ISR.

- Develop and maintain a common defense-wide C4ISR strategic plan;
- Implement a common framework for architecture development for all C4ISR activities;
- Issue updated and integrated C4ISR-related compatibility, interoperability, integration, and security policy directives; and
- Emphasize integrated C4ISR management, and determine the feasibility of implementing a systems integration management-type process.

2. *Identify joint and defense-wide needs:* C4ISR requirements must "reflect the emerging needs of the Unified Command and Joint Task Force . . . Commanders, . . . be flexible enough to accommodate uncertainty," and be fully integrated.

- Increase integration by implementing a standardized, mission-oriented approach to requirements definition using the collaborative Joint Mission Area Assessments and Joint Mission Needs Analyses;
- Create a top-down integrated, nested set of requirements; and
- Apply improved assessment practices (i.e., streamline the existing

Assessment of Alternatives) and implement a simplified, interactive, standardized process for analysis.

3. *Align programs and resources:* The entire DOD portfolio of investments should be assessed and managed from a joint and defense-wide perspective. In addition the "[s]trategic management of C4ISR needs to rely increasingly on incentives for achieving strategic goals . . . and on measures of performance which serve as management controls."

- Strengthen linkages between the Joint Strategic Planning System and other defense-wide requirements processes, and the Planning, Programming and Budgeting System processes at all levels; and
- Align defense resources with joint priorities and requirements.

4. *Expedite the delivery of C4ISR capabilities:* DOD has not applied its new way of doing business (as evidenced by the advanced concept technology demonstrations, Advanced Warfighting Experiments, etc.) "uniformly across the C4ISR arena and has not taken advantage of their potential."

- Consider evolutionary acquisition and other non-traditional acquisition methods for C4ISR; and
- Create a comprehensive management process (i.e., a C4ISR Integrated System Support) to organize ongoing defense-wide C4ISR activities, thereby creating a unified approach to C4ISR system development.

5. *Share knowledge/provide a common infrastructure:* DOD is not fully capitalizing on its investment of information resources and human capital.

- Create a defense-wide C4ISR knowledge base/warehouse with integrated tool sets; and
- Educate, train, retrain, and certify the work force.

Conclusions: In general, the recommendations provided by the Integration Task Force would lead to incremental improvements throughout the DOD. These recommendations are intended to work together to strengthen C4ISR roles and improve the efficacy of C4ISR processes and capabilities.

B.1.2 *1996 Report of the Advanced Battlespace Information System Task Force*

The Advanced Battlespace Information System Task Force was created by the Director, Defense Research and Engineering and the Joint Staff director for C4 Systems "to explore how emerging information and technologies could be used to provide the warfighter with significant new capabilities" identified in *Joint Vision 2010*. The focus of the task force's study was force operations (specifically the concepts of dominant maneuver, precision management, and full-dimensional protection identified in *Joint Vision 2010*), with the C4I portion of the system of systems as the focal point. The task force released its report in May 1996.[2]

Background: The Advanced Battlespace Information System is a set of systems "that forms an underlying grid of flexible, shared, and assured information services and provides advanced capabilities in support of new command and control and force employment concepts." The vision for the Advanced Battlespace Information System is that it will provide a "knowledge-based C4I system environment that facilitates revolutionary operational capability by enabling warfighters to rapidly acquire and use all available information."

Advanced Battlespace Information System Capability Framework: The task force identified an Advanced Battlespace Information System capability framework composed of three tiers—effective force employment, battlespace awareness, and a common information grid—arranged and supported from the bottom up with the information grid providing the infrastructure and services.

New Force Employment Concepts: The doctrine of information superiority espoused in *Joint Vision 2010* will enable commanders to "control and shape the pace and phasing of battle by rapidly integrating and synchronizing dispersed forces to mass effects at the right place and time." In short, the ability to shape the battlefield through information superiority will allow for coordination of force elements to achieve overwhelming effect and attack priority targets. This capability is enhanced by battlefield visualization.

New Command and Control Concepts: The new force employment concepts described above require "a flexible, agile, distributed command structure, with a capability for continual proactive planning and empowered execution." Currently, command and control structures reflect a rigid hierarchy and division of functional areas. New command and control organizations need to be adaptive, and the planning processes need

---

[2]Advanced Battlespace Information System Task Force. 1996. *1996 Report of the Advanced Battlespace Information System Task Force*, Department of Defense, Washington, D.C.

to be dynamic. The Advanced Battlespace Information System architecture supports a decentralized approach that enables distributed empowerment since information superiority allows for "distributing decision making while maintaining coherence across the force."

Mapping Operational Capabilities to Technology Developments: A methodology for mapping operational capabilities to key needed technology developments was developed by the task force, and 32 key functional capabilities were identified for future operations to support desired operational capabilities. This mapping is symmetrical, and the task force "found that in most cases, the same functional capability supported multiple operational capabilities, and typically one operational capability depended on multiple functional capabilities."

Advanced Battlespace Information System Technology Roadmap: The Advanced Battlespace Information System is dependent on advanced information technologies and "a sustained, concerted effort is needed to focus research and operational demonstrations in critical areas" from the near term (1997-2000) through the long term (through 2010). The task force created a technology roadmap that depicts continued developments in current and enabling technologies and fully supporting demonstrations.

Implementation Strategy: The task force noted that "fielding [Advanced Battlespace Information System] capabilities requires incremental insertion, adaptation, and assimilation of new operational concepts and technologies" that are guided "by a single long-term vision and a broad community of participants." The implementation process is "evolutionary and iterative."

Initial Steps Toward the Vision: The task force found that the Advanced Battlespace Information System has "produced substantive near-term benefits." The Advanced Battlespace Information System has "served as a catalyst that stimulated the examination of architectural elements that can be incorporated into a Joint Staff operational architecture to support Joint Vision 2010," and results have been incorporated into defense-wide science and technology planning.

B.1.3 *1998 Joint Warfighting Science and Technology Plan*

The Joint Chief of Staffs, in collaboration with the Office of the Secretary of Defense and the service science and technology executives, identified 10 high-priority, joint warfighting capability objectives, which are updated annually to focus the defense science and technology program. In 1998, the joint warfighting capability objectives are information superiority (which uses C4ISR "to acquire and assimilate information needed to dominate and neutralize adversary forces and effectively employ friendly forces" with near-real-time awareness using a robust C4 network); precision force; combat identification; joint theater missile defense; military operations in urban terrain; joint readiness and logistics and sustainment of strategic systems; force protection/dominant maneuver; electronic combat; chemical/biological warfare defense and protection; countering weapons of mass destruction; and combating terrorism.[3] The joint warfighting capability objectives were augmented this year to be more responsive to the issues identified in the Quadrennial Defense Review. They support the operational concepts of Joint Vision 2010.

B.1.4 DOD Inspector General: *Implementation of the DOD Joint Technical Architecture*

In November 1997, DOD's Inspector General issued an audit report titled *Implementation of the DOD Joint Technical Architecture.*[4] The audit found that DOD does not have "an integrated or coordinated approach to implementing the Joint Technical Architecture."

Background: The Joint Technical Architecture (JTA), which was issued in 1996, "is a minimum set of rules governing the arrangement, interaction, and interdependence of parts to ensure that a conformant system satisfies a specified set of requirements." In short, the Joint Technical Architecture provides minimum standards (which are performance based and primarily commercial) and guidelines for interoperability of all DOD C3I (and C4I) systems, which will be periodically updated and eventually include "all DOD systems that produce, use, or exchange information electronically." The Joint Technical Architecture will be implemented through the Common Operating Environment, which "provides a standard set of common software services, such as data management, communications and graphics through standard application program interfaces." The objective of the Inspector General audit, which was conducted from December 1996 through June 1997, was "to assess DOD programs in implementing information processing standards as a means of achieving systems interoperability."

---

[3]Office of the Secretary of Defense. 1997. *1998 Joint Warfighting Science and Technology Plan*, Department of Defense, Washington, D.C.

[4]Department of Defense Inspector General. 1997. *Implementation of the DOD Joint Technical Architecture,* Report No. 98-023, Department of Defense, Washington, D.C.

Implementation of the Joint Architecture: Due to the minimal planning guidance provided by the Office of the Secretary of Defense, the Joint Technical Architecture implementation plans submitted by 17 major DOD components do not reflect a coordinated or integrated DOD approach to implementation. As such, it is unlikely that DOD's interoperability goals will be met effectively or efficiently. The Joint Technical Architecture was jointly implemented by the Under Secretary of Defense for Acquisition and Technology and the Assistant Secretary of Defense for C3I in August 1996.

Component Implementation Plans: Only half of the DOD components responded to the implementation guidance and, overall, the responses received by the Assistant Secretary of Defense for C3I were "incomplete and inaccurate." In fact, when viewed as a whole, the responses "did not represent a uniform structure and a coordinated implementation strategy . . . [and] generally did not identify the component's priority for JTA implementation, estimated cost, or implementation schedule." The Inspector General believed that the Assistant Secretary of Defense for C3I's lack of an overall DOD perspective in the definition of the integration guidelines was a "serious omission" in the guidance to the components. In addition, the Assistant Secretary of Defense for C3I did not clearly specify who should submit implementation plans. Finally, problems regarding oversight and integration were identified. For example, as of June 1997, there was no formal process to "receive, track, evaluate, or provide feedback on the Component JTA implementation plans" (although a review team is being formed).

Factors Affecting Implementation: The Inspector General identified three factors that could enhance the Joint Technical Architecture implementation process. First, the Defense Information Infrastructure Common Operating Environment "provides a standard platform that mission area applications can be designed to access through standardized application program interfaces . . . [allowing] software developers to concentrate on building mission area applications instead of building duplicative system support service software." Secondly, the DOD can build on the Army implementation experience, including the development of the Army Technical Architecture, which serves as the basis for the Joint Technical Architecture. Finally, the DOD Total Asset Visibility Implementation Plan's establishment of "clusters of capability rather than phasing combat support systems one at a time into the Global Combat Support System . . . could establish a model for cross-Service and cross-functional coordination, which is essential for effective and efficient JTA implementation." Several factors were also identified that could impede the implementation process. First, although the Defense Principal Staff Assistants have been given oversight responsibility, their role in implementing the Joint Technical Architecture is not defined. Another factor is the DOD mandate for use of COTS technology, which may not be complementary to the Joint Technical Architecture since "all commercial products may not be built to the standards specified in the JTA" (additionally, there is not a clear method by which to certify commercial software products as Joint Technical Architecture compliant). Finally, although an integrated architecture—consisting of technical, operational, and systems components—is very important, the focus to date has been on the technical architecture. In addition, other factors, such as the Clinger-Cohen Act of 1996, affect the implementation of the Joint Technical Architecture.

Information Technology Management Reforms: The Inspector General's report reviewed the Clinger-Cohen Act of 1996, which requires the DOD "to establish a process to select, manage, and evaluate the results of information technology investments" and to designate a chief information officer. The Assistant Secretary of Defense for C3I is the primary chief information officer for the DOD. In addition, the DOD also established the Chief Information Officer Council to advise on matters related to information technology and coordinate the implementation of the mandates of the 1996 Act.

Conclusion: The Joint Technical Architecture is the key initiative to achieving DOD's goal of interoperability. The Inspector General's review of the DOD component implementation plans indicates "that the JTA is being implemented in an environment that is not consistent with attaining interoperable information processing systems in an integrated and coordinated manner." According to the Inspector General's report, the Office of the Secretary of Defense needs to assume responsibility for establishing "a framework of strategic planning, policy and guidance to support those plans." Additionally, no mechanism has been identified to provide the guidance and oversight to the components that is needed to ensure efficient and coordinated implementation of the Joint Technical Architecture.

Recommendations: The Inspector General provided four recommendations to the co-chairs of the DOD Architecture Coordination Council (i.e., the Under Secretary of Defense for Acquisition and Technology, the Assistant Secretary of Defense for C3I, and the Director of the Joint Staff C4 Systems Directorate):

- Develop a methodology for cross-service and cross-functional coordination of DOD component Joint Technical Architecture implementation plans.

• Develop a methodology to measure and track the progress and success of the Joint Technical Architecture implementation.
• Disseminate information that could enhance or impede implementation of the Joint Technical Architecture.
• Establish review mechanisms to periodically assess joint interoperability levels.

Management Comments: In short, the stakeholders generally concurred with the findings and conclusions of the Inspector General's report with comments, some of which were incorporated in the Inspector General's final report for accuracy and clarification. In addition, they also fully concurred with the recommendations.

B.1.5 The "C4ISR Mission Assessment"

The Assistant Secretary of Defense for C3I and the Joint Staff C4 Systems Directorate co-sponsored the C4ISR mission assessment to address potential C4ISR issues as they relate to the support of DOD's evolving operational concepts and future force and weapons mixes. The resulting "C4ISR Mission Assessment" document provided input to the Quadrennial Defense Review's Modernization Panel. The "C4ISR Mission Assessment" was composed of a number of focused analyses on architecture; C3; communications; intelligence, surveillance, and reconnaissance; mission analysis; and Concept of Operations-Enabled, "which were closely coupled to provide an integrated set of assessments and recommendations across the breadth of the C4ISR domain." A formal C4ISR mission assessment report was never published. The "C4ISR Mission Assessment" document provides a summary of the Communications Mix analysis.

Study Background: The objectives of the Communications Mix study were to examine the adequacy of DOD's communications capabilities; develop alternative investment strategies; assess performance, cost, and risk of alternatives; and recommend an investment strategy, but not specific system designs. The study was limited to an operational scenario consisting of two major conventional theater wars allowing for a pre-positioned posture. It focused on the communications needed to support the operations of the deployed warfighter in three main areas: theater forward communications, tactical wide-area networking, and theater reach-back. The study team reviewed communications requirements data collected from past studies and developed a C4ISR Mission Assessment communications requirements information flow model. Observations derived from the model clearly indicated that information management is critically needed to contain the growth of communications requirements.

As a baseline, the study team examined current and projected spending articulated in the Future Years Defense Plan FY98-03. Of the $257 billion total C4ISR funding projected for this period, approximately $36.8 billion is allocated for communications (excluding intelligence, surveillance, and reconnaissance-funded communications). The systems examined by the study—satellite communications, tactical radios, tactical wide area networks, and long-haul systems, such as the Defense Information Systems Network—account for $16.1 billion, or 44%, of DOD's communications spending over the period.

Assessment: The Communications Mix study assessed the identified systems to determine the ability of DOD's current and programmed communications systems to meet projected future requirements. As such, it was determined that today's "communications systems supporting the deployed warfighter are currently able to support only a fraction of projected future data rate requirements. In broad terms, the magnitude of the shortfall ranges from a factor of four . . . at the upper echelons to a factor of fifty . . . at the tactical radio level." Therefore, communications capabilities must be "increased dramatically" to support projected requirements. The study also considered alternatives to address the shortfalls identified in communications capabilities.

Recommendations: The "C4ISR Mission Assessment" recommended the following changes to the current communications portfolio to address the deficiencies identified in the areas of tactical radios, joint tactical wide area networks, joint network and services management, military satellite and fiber communications, commercial leases, and unmanned aerial vehicles communications relays:

- "Accelerate the procurement of the next generation wide-band military satellite system to address the shortfall in available capacity for theater reachback and intra-theater long haul communications."
- "Accelerate and coordinate service programs for upgrades to the communications switching and trunking systems supporting the deployed tactical terrestrial [WAN]."
- "Develop and procure a Joint Tactical Network and Services Management capability and develop the necessary concept-of-operations and procedures for dynamically monitoring and managing communications assets."
- "Consolidate the Services' multi-band, multi-mode radio . . . programs and develop a family of programmable, modular digital radios based on a common modular radio architecture."
- "Procure two squadrons of five . . . UAV communications relay

aircraft each to provide early entry and surge communications relay capability."

- "Initiate R&D to provide high-data rate protected communications services with a combination of satellites and UAV relays."
- "Initiate demonstration programs to assess the utility of emerging commercial mobile subscriber services and technology to augment or replace the existing UHF satellite communications systems."

In addition, specific recommendations for deployed warfighter communications were also made in each area. The actions recommended by the C4ISR Mission Assessment study team would add an estimated $5 billion to the Future Years Defense Plan. Finally, five investment options were provided to decision makers for "program actions that provide increasing levels of capability at increasing levels of investment."

B.2 DEFENSE SCIENCE BOARD STUDIES

B.2.1 Defense Science Board Task Force on C4ISR Integration

In 1995, the Defense Science Board established the task force on C4ISR Integration at the request of the Under Secretary of Defense for Acquisition and Technology as part of DOD's attempt to accelerate the development of C4ISR integration and architecture efforts. The task force was charged with providing advice to the chair of the Integration Task Force (ITF) on all aspects of C4ISR as well as preparing separate reports of its judgments on C4ISR issues. The task force released its report in February 1997.[5]

Background: After meeting with the Integration Task Force to hear about its process, organization, and results to date, the Defense Science Board task force formulated a set of recommendations to be considered by the Integration Task Force and submitted two letter reports. The Defense Science Board task force found that DOD's "ITF efforts were overly broad and complex" and that it was difficult to accomplish the tasks defined by the ITF because of the "fractionated and 'stovepiped' nature of the C4ISR stakeholder community, particularly in regard to programmatic and fiscal responsibilities." The task force also concluded that the Integration

---

[5]Defense Science Board. 1997. *Report of the Defense Science Board Task Force on Command, Control, Communications, Computers, Intelligence, Surveillance and Reconnaissance (C4ISR) Integration,* Office of the Under Secretary of Defense for Acquisition and Technology, Washington, D.C.

Task Force's recommendations spoke to a "generalized Pentagon process" that would not "result in a leveraged process in achieving important new levels of C4ISR integration."

The Defense Science Board task force was concerned about the lack of a process for combining the services' C4ISR equipment and procedures. Although DOD created various joint committees to address the related issues, the task force did not believe they were "adequate to deal with the joint C4ISR problem." It concluded that "the fundamental responsibility" belonged to the Chairman of the Joint Chiefs of Staff and the unified and specified commanders-in-chief (CINCs). As such, the task force identified two needs—"improving the joint process for determining what a joint force commander needs in order to operate effectively, and the creation of a joint system engineering organization"—that are described below.

Joint Process: The Defense Science Board task force envisioned a "more formal joint process on the front end of the programming and budgeting cycle that gives joint force commanders stronger influence on decisions regarding what increased (or decreased) capabilities are needed for them to carry out their assigned missions." The task force, therefore, recommended customer-based, output-oriented planning and programming in which the joint operational customer has a formal role in "formulating joint operational concepts and . . . architectures, as well as ensuring appropriate input to resource allocation priorities to produce effective joint operational forces." It also defined different roles for DOD's three C4ISR integration communities—the Joint World (e.g., the Chairman of the Joint Chiefs of Staff), the Office of the Secretary of Defense, and the military departments and agencies. In addition, the joint customer would play a leading role in motivating a shift of resources from support infrastructure to operational, or forcing, capabilities. The task force was concerned that more than half the defense budget was allocated for support infrastructure, allowing for a "critical imbalance."

The task force recommended expanding the joint role in the planning and budgeting process by "insuring that the joint elements of the Department fulfills [sic] their responsibilities and that the joint operational needs become paramount from the outset." As such, the role of the CINCs needs to "become a more integral and required part of the process." They should be treated as the customers and the process "should evaluate results based on satisfying customer's needs." Joint planning and programming should focus more on "providing the right set of a capabilities for the CINCs to carry out their operational missions." In addition, the CINCs would have substantial influence on the Chairman of the Joint Chiefs of Staff input to Defense Plans and the services' plan of the month process based on their input "on gaps in their capability to meet assigned mission needs."

Finally, the task force found that there is no effective process for providing guidance for joint operational doctrine and architectures as they relate to the development of the connectivity required for an effective joint force C4ISR integration, which could compromise the "ability to respond rapidly with effective joint forces." As such, a process would be required "to develop joint operational doctrine with enough specificity to guide joint operational architectures," which, in turn, must be specific enough to guide the system and technical architectures. They stated that "doctrine and architectures must fill the twin needs of adaptability to CINC unique needs and structuring deployable capabilities to fit a variety of CINC needs." The Chairman of the Joint Chiefs of Staff/Joint Staff and U.S. Atlantic Command should share the lead for developing joint operational doctrine and architectures, and the "key implementing principle must be that the CINC's part of the front end process become an essential prerequisite to the follow-on planning and budgeting."

Joint Systems Engineering Organization: As mentioned above, the task force found that there is a "need for a 'military engineering organization to support CJCS and the CINCs in their role in joint C4ISR.'" It also identified eight functions for the Chairman of the Joint Chiefs of Staff and CINCs in order to carry out their responsibilities for the design of the joint operational architecture. The task force further defined the organization of and resources for a military systems engineering capability for C4ISR integration. The estimated cost of such a capability was approximately $50 million per year. In addition to the creation of such an organization "to support the CINCs in their evolving responsibility for the operational design of joint C4ISR," the task force recommended "that the CJCS use the new structure that was established to provide joint operational architectures and joint system engineering to Joint Theater and Air and Missile Defense as a pilot program for the broader C4ISR area, with focus on the refining [of] the responsibilities and missions of warfighting CINCs."

Other Issues: Several other issues were identified by the task force regarding DOD's management of C4ISR integration:

- *Intelligence support to military operations.* The task force recommended that DOD "work with the DCI [Director of Central Intelligence] and the broader Intelligence Community to develop new ways of providing information support for operational commanders which effectively and efficiently integrate the rich array of assets available within the United States."
- *Vulnerability, security, and protection.* The task force recommended that the DOD "should closely evaluate whether the separation of intelli-

gence and operations within warfighting elements continues to serve the nation well."

- *Acquisition of C4ISR capabilities.* The task force identified two "unique characteristics of C4ISR systems relevant to the acquisition process. First, the inherently joint aspects of C4ISR are critical to the overall utility of C4ISR . . . . The second key characteristic is the pace of technological change in the field of information systems that form the basis for much of C4ISR," which is "totally incompatible with normal DOD procurement practices." As such, DOD needed "to push harder on acquisition reform."

B.2.2 Improved Application of Intelligence to the Battlefield

In February 1997, the Defense Science Board task force on Improved Application of Intelligence to the Battlefield (May–June 1996) released its report extending the 1995 work on the same topic.[6] It should be noted that this follow-up study was conducted and the report drafted on the eve of the Bosnian elections in September 1996.

Background: The 1996 task force was directed to "review the progress towards the implementation of recommendations made" in 1995 and "to determine any improvements which would enhance the flow of intelligence and other information for Operation Joint Endeavor," with an emphasis on other C4SIR improvements that could be quickly applied to support coalition forces as well as future operations after the restructuring and redeployment of forces, especially ground forces, in December 1996. It should be noted that all of the recommendations resulting from the 1995 study, which addressed policy, technological, and organizational deficiencies that would affected the safety of U.S. forces, were accepted by the DOD and Central Intelligence Agency, and "approximately $150 million followed to begin making rapid improvements centered primarily around Air Force and Navy missions."

Key Findings and Recommendations: The 1996 task force found that the findings from the 1995 study were being implemented effectively and that there was a "dramatic improvement in information availability to the forces." In addition, the task force made recommendations in four major areas to extend the progress achieved since the 1995 study:

- *Information integration.* The 1996 task force found a critical need to integrate combat and information power to better match information ca-

---

[6]Defense Science Board. 1997. *Report of the Defense Science Board Task Force on Improved Application of Intelligence to the Battlefield (May-June 1996)*, Office of the Under Secretary of Defense for Acquisition and Technology, Washington, D.C.

pability with mission requirements and to provide more information and better connectivity.

- *Joint Broadcast System Advanced Concept Technology Demonstration (Bosnian Command and Control Augmentation).* The 1995 task force found that insufficient bandwidth and poor imagery quality were a problem for both U.S. and coalition operators, and the Bosnian Command and Control Augmentation was designed as a remedy to provide "relevant, timely information (specifically large data format information such as imagery and video)" to these operators. In this major area, the 1996 task force made specific recommendations on providing additional time and funding for the Bosnian Command and Control Augmentation, which is not an official advanced concept technology demonstration; providing greater information support that is required for brigade and battalion headquarters; and addressing information management challenges.
- *Leave-behind programs.* The 1996 task force recommended that an interagency task force be established "to identify opportunities, develop specific items, and assist in deployment before redeployment phase."
- *Areas for other major recommendations.* The 1996 task force also made recommendations in 11 other areas: C4ISR dynamic tasking capability (in short, providing tools and processes "to dynamically integrate tasking of national/theater [reconnaissance]/surveillance in C2 systems with timely feedback"; the task force found "that a failure to coordinate and integrate the use of superb ISR assets in direct support of the warfighter is a remaining barrier to achieving and exploiting information dominance"), human intelligence information management, countermine/demining, Linked Operations-Intelligence Centers Europe, airborne video surveillance, tactical signal intelligence, commercial equipment, Joint Surveillance Target Attack Radar System, commercial satellite imagery, information warfare vulnerability, and total asset visibility. It also identified recommendations made in 1995 that required renewed attention, focusing on Bosnia theater radar/infrared imagery, controlled imagery base, ultra-high-frequency satellite communications, hard copy, linguists, and communications landing rights.

The 1996 task force also uncovered some "great ideas": the DOD and military were adapting to the changing environment, as evidenced in the shift of missions in Bosnia; important information applications were developed; there were signs of effective information integration; and there were innovative uses of information.

B.2.3 *Tactics and Techniques for 21st Century Military Superiority,* 1996

In 1996, the Defense Science Board summer study task force examined innovative tactics for improving the effectiveness of rapidly de-

APPENDIX B   265

ployable forces with regard to future warfighting capabilities. The Defense Science Board released its final report and two accompanying volumes of supporting materials and white papers in October 1996.[7]

Background: The 1996 Defense Science Board task force was asked to identify how to make small and rapidly deployable forces more effective with the goal of accomplishing "missions heretofore only possible with much larger and massed forces." The task force considered the scenarios posited by the 1995 Defense Science Board summer study, which determined that, in the future, adversaries will have the motive and means (through advanced technologies) to achieve military superiority. As a result, it was determined that the United States must increase the effectiveness and decrease the vulnerabilities of rapidly deployed forces to enhance its "freedom of action to deal with this future."

New Expeditionary Force Concept: The task force defined goals for a new joint expeditionary force (or leading-edge strike force) that focus on massing fires rather than forces. This new force would be composed of "light and agile ground and air combat cells coupled to remote suites of sensors, weapons, and information processors." The size and composition of the combat cells would be determined by the nature of the mission. These forces would be distributed and disaggregated, empowered by unprecedented situational understanding (which is a higher level of knowledge than situational awareness), dependent on remote fires that are effective against a variety of targets, connected by a robust information infrastructure, and supported by precision logistics.

Operational Considerations: Two factors would remain constant in any operating environment, regardless of force size and composition: "dependence on remote elements and ground forces organized around agile combat cells." In general, remote strikes using air and naval forces would precede deployment of ground units. Then, an initial small intensive force would be inserted (this force could be either concentrated to coordinate security or distributed to increase survivability and enlarge territorial control, depending on the circumstances). To achieve dominant situational understanding, the task force envisions a multilayered sensor approach integrating surveillance and connectivity, which "would enable effective remote fires and militarily useful combat cell operation." In order to free more lift resources for combat operations, the task force recommends reducing the support functions deployed to theater operations. The C4ISR

---

[7]Defense Science Board. 1996. *Tactics and Techniques for 21st Century Military Superiority*, Office of the Secretary of Defense, Washington, D.C.

infrastructure, which could be effectively deployed remotely through the information infrastructure, is identified as a candidate for reduction.

Analyses and Simulation: In order to examine its new expeditionary force concept, the task force sponsored several analyses and simulations, which are reviewed in Section IV of the final report.

Enabling Elements of Concept: The new expeditionary force concept depends on the synergies and the interdependency between the following functions/capabilities: remote fires; battle management, command and control; information infrastructure; situation understanding; protection and survivability of ground forces; and training. In discussing the importance of battle management, command and control, the task force breaks down C4ISR into two interdependent categories—the human function of command and the technical function of the C3ISR activities—and emphasizes the "need to maintain human relationships on [the] dispersed, digital battlefield."

Recommendations: The task force offered three sets of recommendations for the Secretary of Defense and the Chair of the Joint Chiefs:

  • *Establish a joint effort to explore and evolve this new force concept.* The task force calls this its "try before buy" recommendation, and calls for testing and analysis as well as augmenting activities that are emerging within the services, such as the Army After Next initiative. This would be supported by redirected analysis and simulation activities, and an executor (or executive agent) would be selected to lead the effort and evolve the concepts.
  • *Support critical and enabling systems and mechanisms* by accelerating the development of the information infrastructure architectures. A joint warfighter, or operational, architecture should be developed that addresses operations concepts; processes and procedures for information generation, condition, fusing, and use; weapons, sensor, and platform functional characteristics; assignment of functions; and force structure. A joint technical information architecture should be mandated by the Under Secretary of Defense for Acquisition and Technology and the Assistant Secretary of Defense for C3I that addresses coherent data formats, protocols, message standards, interfaces and so on; enables open systems; and provides a "building code" for the information architecture. Finally, a joint information infrastructure systems architecture should be implemented by the services' C4I organizations that migrates legacy systems and integrates commercial systems. In addition, the task force calls for supporting both existing and candidate advanced concept technology

demonstrations and advanced technology demonstrations important to the concept as well as initiating new ones.

• *Prepare to establish a Joint Expeditionary Task Force by 1998 to be the focal point for transitioning the concepts.* This joint operational force would be established under the U.S. Atlantic Command and is envisioned to test the products (that is, the tactics and technologies) of these efforts described by the summer study task force.

Conclusions: The task force believes that there are several necessary conditions already in place for the new capabilities it envisions: there is a compelling strategic rationale; the enabling technologies are maturing rapidly; and there are efforts currently under way in the services to explore these new concepts. In addition, the task force believed that the concepts identified can be "refined, tested, modified, shaped, and evolved into field capabilities over the next 10-20 years." Finally, four complementary concept enablers were identified: fielding the robust information infrastructure; turning situational awareness into situational understanding by managing sensors and information in conceptual contexts; making remote fires work; and operating in a disperse posture.

B.3 GENERAL ACCOUNTING OFFICE STUDIES

B.3.1 *Joint Military Operations: DOD's Renewed Emphasis on Interoperability Is Important But Not Adequate*

In 1994, GAO issued a report on DOD's C4I system and operational interoperability as a follow-up to a 1987 report that identified problems in this area as related to C3 systems.[8] At the time this report was released, the General Accounting Office determined that DOD's success in achieving interoperability during joint operations would be "highly dependent on the availability of a comprehensive, integrated, and useful C4I architecture."

Background: The General Accounting Office found that problems associated with interoperability were persistent, as identified by several reports issued by DOD and the Joint Staff. Cited was the DOD's joint tactical C3I architecture, which was a series of functional area documents published between 1988 and 1992 that "identified service missions, roles and responsibilities; command and control connectivity requirements; and support-

---

[8]General Accounting Office. 1993. *Joint Military Operations: DOD's Renewed Emphasis on Interoperability Is Important But Not Adequate*, General Accounting Office, Washington, D.C.

ing C3 systems and equipment." According to the General Accounting Office, DOD representatives did not consider this a useful planning document even though a number of system and operational interoperability problems were identified. Interoperability (system, technical, and operational) was also addressed in 1991 by a panel formed by the Joint Chiefs (see the *Command and Control Functional Analysis and Consolidation Review Panel Report*). In 1992, another Joint Staff team looked at interoperability as it related to C2 systems. A third report cited was DOD's 1992 report to Congress on the Persian Gulf War. In this report, DOD cited interoperability problems identified during the joint operations and the challenges that remain ahead.

C4I for the Warrior: This initiative, launched in 1992, intended "to (1) address joint force C4I interoperability issues and (2) provide a means for unifying the many heterogeneous service C4I programs." The General Accounting Office found that achieving this initiative would be a prolonged process due to its three concurrent—quick-fix, mid-term, and objective—phases. In addition, the General Accounting Office concluded that a comprehensive architecture remained to be developed, despite the DOD's joint tactical C3I architecture (see comments above). The General Accounting Office also found that the Joint Interoperability and Engineering Organization, which was responsible for the architecture, "lacked the authority to enforce compliance with interoperability standards." Finally, there was a continuing concern regarding effective interoperability enforcement despite DOD efforts to strengthen enforcement of C4I interoperability.

The General Accounting Office concluded that DOD had the means to strengthen C4I interoperability. In 1993, the Secretary of Defense directed that the U.S. Atlantic Command assume a new mission as a joint headquarters for U.S.-based forces, based on a recommendation from the Chairman of the Joint Chiefs of Staff. The General Accounting Office concluded that the Command would be "ideally suited for additional responsibilities associated with C4I interoperability. Specifically, the Command could be assigned primary responsibility for assessing C4I requirements for the potential effect on joint force operations." The Command could also advise the Defense Information Systems Agency on the development of the joint C4I architecture as well as ensure "continuous C4I interoperability assessments through joint training exercises."

Recommendations: The General Accounting Office identified three areas to assist DOD's ability to achieve C4I interoperability:

- The provision of guidelines for developing the joint C4I architecture, including time-driven goals.

APPENDIX B   269

- The establishment of a joint program management office "with directive authority and funding control for C4I systems acquisitions."
- The consideration of assigning responsibility to the U.S. Atlantic Command for C4I interoperability, as described above.

DOD Response: Although DOD generally agreed with the report's findings, it believed it had taken "adequate measures to deal with C4I system interoperability and saw no benefit in assigning additional responsibilities to the U.S. Atlantic Command."

B.3.2 Joint Military Operations: Weaknesses in DOD's Process for Certifying C4I Systems' Interoperability

In March 1998, the General Accounting Office completed its review of the certification process for interoperability of C4I systems and concluded that DOD stakeholders (CINCs, the services, and the DOD agencies) are generally not complying with the C4I certification requirement.[9]

Background: In 1992, the Defense Information Systems Agency (DISA) established a certification process to ensure interoperability of C4I systems during joint operations as a result of interoperability problems experienced during the Persian Gulf War. This process tests and certifies existing, newly fielded, and modified systems for interoperability. New systems are generally denied approval for production if they have not been certified. Although a system may pass certification testing, it is possible that it has not been tested against all systems with which it may be interoperable. Finally, a waiver may be granted to drop certification requirements for "developmental efforts, demonstrations, exercises, or normal operations"; however, this is not a permanent waiver and typically is granted for 1 year. The Joint Staff's director for C4 systems is responsible for ensuring compliance, and the Defense Information Systems Agency's Test Command is the sole certifier of systems.

Findings: The GAO found that DOD's compliance with the certification requirement is inadequate:

- Test Command analysis indicates that "a significant number of existing C4I systems had not been submitted for certification testing," and is

---

[9]General Accounting Office. 1998. *Joint Military Operations: Weaknesses in DOD's Process for Certifying C4I Systems' Interoperability*, Report No. NSIAD-98-73, General Accounting Office, Washington, D.C.

unable to identify how many systems actually require certification (e.g., some systems are legacy systems or stand-alone systems). During FY1994-1997, only 149 systems were certified by the Test Command.[10]

- No newly developed systems of the C2 Initiatives Program were certified and, for the past 3 years, no advanced concept technology demonstrations were tested or certified. Finally, there is no consistency with regard to recertification of modified systems.
- There are several reasons for inadequate compliance: lack of knowledge of the certification requirement by system managers (although some managers purposely did not submit their fielded or modified systems for testing); inadequate budgeting by the services for the testing and certification process; and production approval for some new systems without verification of the certification process.

The GAO also found weaknesses in DOD's certification process:

- The Test Command does not have a way to focus its limited resources on certifying crucial systems because a "complete and accurate listing of C4I systems requiring certification and a plan to prioritize systems for testing" does not exist. For example, of the 42 existing C2 systems submitted by the services and determined to be crucial to military commanders by the Military Communications Electronics Board, 23 had not been tested or certified.
- The Test Command does not advise the services about interoperability problems observed during joint exercises. During the four joint exercises held between 1996 and 1997, "the Test Command noted that 15 systems experienced 43 'significant interoperability problems'—defects that could result in loss of life, equipment, or supplies"—most of which were caused by system-specific software problems. If the services are not notified of these problems, "significant interoperability problems may arise in subsequent exercises and operations." It should be noted, however, that Test Command officials are looking at ways to formally track and follow up on these problems.
- The Test Panel does not have a formal process for informing DOD stakeholders about expired waivers.

---

[10]The Defense Information Systems Agency's Defense Integration Support Tool database of C4I systems listed about 1000 systems that may exchange information with other systems, and there are approximately 1176 unclassified intelligence systems as well. In addition, the Defense Integration Support Tool (which GAO has reported to be inaccurate and incomplete) only recently included certification status as part of the database.

APPENDIX B 271

Recommendations: The General Accounting Office made the following recommendations:

- To make sure that critical systems do not proceed into production without consideration given to the certification requirement, the Secretary of Defense should "require the acquisition authorities to adhere to the requirement that C4I systems are tested and certified for interoperability prior to the production and fielding decision unless an official waiver has been granted."
- To improve the interoperability certification process, the Secretary of Defense, with advice from the Joint Chiefs, should direct the services to review the information in the Defense Integration Support Tool for verification and validation and compile a complete listing of all C4I systems that require certification. In addition, the Defense Information Systems Agency director should ensure that the status of a system's certification is incorporated into the Defense Integration Support Tool and that this database is "properly maintained to better monitor C4 systems for interoperability compliance."
- The Secretary of Defense should request that the chairman of the Joint Chiefs direct the Joint Staff, in collaboration with the DOD stakeholders, to develop processes (1) to prioritize C4I systems for testing and certification and (2) to formally follow up on and report to the stakeholders interoperability problems identified during joint exercises and inform stakeholders of systems that require interoperability testing.
- Finally, a system to monitor waivers should be established by the Chairman of the Joint Chiefs of Staff.

In addition, the report provided an appendix that briefly reviews the DOD initiatives currently under way that address aspects of interoperability: the C4I for the warrior concept; the C4ISR Architecture Framework; the Defense Information Infrastructure strategy; and the Levels of Information Systems Interoperability initiative.

Response: DOD generally concurred with the General Accounting Office findings and was firmly committed to improving its interoperability certification process by taking action to implement the General Accounting Office's recommendations.

Appendix C

Members of the Committee

James C. McGroddy, *Chair,* was a senior vice president at IBM until his retirement at the end of 1996. He is chairman of the board of Integrated Surgical Systems, a major player in the medical robotics field. He also serves as an advisor to several government agencies, as a member of a number of National Research Council panels, and as a visitor and advisor at several universities. As senior vice president, IBM Research, from 1989 to the end of 1995, he was responsible for the work of about 2500 technical professionals in seven research laboratories around the world. Two of these laboratories, in Beijing, China, and in Austin, Texas, were established under his leadership. He was also a member of IBM's Worldwide Management Council and its Corporate Technical Committee. Dr. McGroddy originally joined IBM in its Research Division in 1965 after receiving a Ph.D. in physics from the University of Maryland. He earned his B.S. in physics from St. Joseph's University in Philadelphia in 1958. In his first years at IBM Research he focused on research in solid state physics and electronic devices, and as a result of achievements in these areas was named a fellow of both the Institute of Electrical and Electronic Engineers (IEEE) and the American Physical Society. In the 1970-1971 academic year he was a visiting professor of physics at the Danish Technical University. Returning to IBM, he served in a number of management positions in research, development, and manufacturing before returning to head the Research Division in 1989. He is a member of the National Academy of Engineering; chairman of the board of trustees at Phelps Memorial Hospital Center in Sleepy Hollow, New York; and a trustee of the HealthStar Hospital Network, of the Guglielmo Marconi Foundation, and of St. Joseph's University. He also serves on the board of directors of the Paxar Corporation.

Charles Herzfeld, *Vice Chair*, currently serves as a consultant to a variety of organizations, such as the Defense Advanced Research Projects Agency, Los Alamos National Laboratory, and others. He holds an engineering degree from the Catholic University of America (B.S., 1945) and a Ph.D. from the University of Chicago (1951). He worked as a physicist at the Ballistic Research Laboratory, Aberdeen, Maryland, from 1951 to 1953, and at the Naval Research Laboratory in Washington, D.C., from 1953 to 1955. After several years with the National Bureau of Standards, he became assistant director of the Advanced Research Projects Agency of the Department of Defense. He was director of the Defense Advanced Research Projects Agency from 1965 to 1967 and was instrumental in setting up the ARPANET. During his several years of affiliation with ITT, Dr. Herzfeld served as technical director and director of research groups and finally as vice president, director of research (1979-1983) and director of research and technology (1983-1985). He served as director of Defense Research and Engineering at the Department of Defense (1990-1991) and was a consultant to the Office of Science and Technology Policy, Executive Office of the President, in 1991. Dr. Herzfeld received the Flemming award in 1963 and was awarded the Meritorious Civilian Service medal by the Department of Defense in 1967. He has contributed numerous articles to professional journals.

Norman Abramson is vice president and chief technical officer of ALOHA Networks, a San Francisco company providing satellite access to the Internet using small Earth stations. He joined the Stanford faculty in 1958 as assistant and then associate professor of electrical engineering. In 1965 he was appointed professor of electrical engineering at the University of Hawaii. He also served as professor and chairman of the Computer Science Department at the University of Hawaii. In 1967 he assumed the position of director of the ALOHA System, a university research project concerned with new forms of data network architecture. From 1972 to 1985 he served as a United Nations adviser to developing countries on the use of satellite technology for national development. In 1995 he left the University of Hawaii to found ALOHA Networks Inc. in order to develop advanced forms of ALOHA channels in the commercial sector. In addition to his fundamental research in multiple access communications, Mr. Abramson directed the creation of the ALOHANET, a wireless packet network operating throughout Hawaii.

Edward Balkovich is a director at Bell Atlantic. He is responsible for IP and data network system engineering in Bell Atlantic's Network Architecture organization. He also contributes to the technology adoption strategy and network evolution plan. Most recently, he led the introduction of voice over IP services in Bell Atlantic's core network. Dr. Balkovich's areas of expertise include computer-based systems and networks. Before coming to Bell Atlantic, Dr. Balkovich was senior consulting engineer with Digital Equipment Corporation. At Digital, he was responsible for a variety of research, architecture, and integration activities, and was a technical partner to major corporate accounts. While at Digital he co-led Project Athena at Massachusetts Institute of Technology and contributed to the architecture of the VAX cluster product line and the demonstration of encryption, tunneling, and firewalls as the basis for secure use of the Internet. Dr. Balkovich has also held a number of academic appointments, including adjunct associate professor at Brandeis University, visiting scientist at the Massachusetts Institute of Technology, and assistant professor at the University of Connecticut. Dr. Balkovich received his B.A. in mathematics from the University of California at Berkeley, and his M.S. and Ph.D. in electrical engineering and computer science from the University of California at Santa Barbara. He is a member of IEEE and the Association for Computing Machinery.

Jordan Baruch received a B.S. and an M.S. in electrical engineering (1948) and an Sc.D. in electrical instrumentation (1950) from the Massachusetts Institute of Technology and served as an assistant professor and lecturer in electrical engineering until 1970. Dr. Baruch has been president of Jordan Baruch Associates in Washington, D.C., since 1981. He is a consultant to industry and government on the planning, management, and integration of strategy and technology. Previously he was general manager, Medinet Department, General Electric Co. (1966-1968); president, Educom (1968-1970); independent consultant (1970-1971); lecturer in business administration, Harvard University (1971-1974); professor, Tuck School of Business and Thayer School of Engineering, Dartmouth College (1974-1977); and assistant secretary for science and technology, U.S. Department of Commerce (1977-1981). Dr. Baruch is a member of the National Academy of Engineering, the IEEE, and the American Association for the Advancement of Science. His research interests include computers in communication, acoustics, and technology management.

Richard J. Baseil is vice president in Telcordia Technologies' Professional Services organization. Mr. Baseil has managed product testing and quality analyses of telecommunications switching, signaling, transport, and customer-premise systems, with an emphasis on hardware and software interoperability. He also advises telecommunications service providers on improvements to their procurement processes for network systems.

Baseil played a major role in defining the industry need for, and subsequently establishing, a multi-company internetwork interoperability test planning effort in the United States, and he managed the Bellcore staff and the interconnection facility used by industry participants to conduct nationwide signaling and interoperability testing. Mr. Baseil has 24 years of telecommunications experience, having had responsibility for switching systems engineering, signaling network engineering, operations systems engineering, operating services system requirements, network database requirements, ISDN data services engineering, billing services, and some early descriptive work on next-generation switching systems. Mr. Baseil holds bachelor's and master's degrees in electrical engineering from the New Jersey Institute of Technology.

Thomas A. Berson is founder and president of Anagram Laboratories, a company that specializes in computer security and cryptography. Dr. Berson has deep knowledge of cryptosystem architecture, cryptographic algorithms and protocols, network security issues, tiger team analyses, and strategies for information conflict. His consulting practice is focused on market-leading multinationals and U.S. government agencies. He earned a Ph.D. in computer science from the University of London and a B.S. in physics from the State University of New York. He has been a visiting fellow in mathematics at the University of Cambridge and is a member of the Stanford University Cryptography Seminar. He is an editor of the *Journal of Cryptology*. He is past-president of the International Association for Cryptologic Research and is the incoming chair of the IEEE Technical Committee on Security and Privacy. Toward the end of this study, Dr. Berson was appointed principal scientist at the Xerox Palo Alto Research Center.

Richard Kemmerer is professor and past chair in the Computer Science Department at the University of California at Santa Barbara. He is a nationally known consultant in computer security and formal verification. He has written widely on the subjects of computer security, formal specification and verification, software testing, programming languages, and software complexity measures. Dr. Kemmerer received a Ph.D. (1979) in computer science from the University of California at Los Angeles. He is a fellow of the IEEE Computer Society and of the Association for Computing Machinery and past chair of the IEEE Technical Committee on Security and Privacy. He also served on the National Bureau of Standards' Computer and Telecommunications Security Council and on the National Research Council's study committees that produced *Computers at Risk* and *For the Record*.

Butler Lampson is an engineer with the Microsoft Corporation. He was previously a corporate consulting engineer for Systems Research Center, Digital Equipment Corporation. Dr. Lampson has several publications and patents to his credit. He is a member of the Association for Computing Machinery, International Federation for Information Processing Working Group 2.3 on Programming Methodology, and the National Academy of Engineering. He received his Ph.D. (1967) in electrical engineering and computer science from the University of California and his AB magna cum laude (1964) with highest honors in physics from Harvard University.

David M. Maddox retired from the U.S. Army in 1995 after serving as Commander in Chief, U.S. Army in Europe. Since that time, he has performed extensive consulting services regarding concepts, systems requirements, analytic techniques and analyses, operations and systems effectiveness, and program capture strategies to civilian corporations, government agencies, and defense industries. General Maddox has had extensive command experience. He served four tours in Germany during which he commanded at every level from the platoon through the Army group and theater. After commanding at the platoon and troop level in the 14th Armored Cavalry Regiment, he later commanded the 1st Squadron, 11th Armored Cavalry Regiment in Fulda, the 2nd Armored Cavalry Regiment (he was the 61st Colonel of the Regiment) in Nuremberg, the 18th Infantry Division (mechanized) in Bad Kreuznach, V Corps in Frankfurt, and NATO's Central Army Group and U.S. Army, Europe and 7th Army in Heidelberg. In addition, he has significant background in operations research.

Paul D. Miller is chairman and CEO of Alliant Techsystems. Admiral Miller has had extensive command experience. He retired in November 1994 as Commander in Chief of the U.S. Atlantic Fleet (CINCLANT) and Supreme Allied Commander (Atlantic) for NATO. As CINCLANT, he oversaw the execution of Operation Uphold Democracy in Haiti. In his active service, he was a strong advocate for joint and combined operations. He developed a reputation as an innovator in the use of new technologies to support military operations. Among other notable accomplishments, he led the reorganization of the U.S. Atlantic Command, the first command that integrated all combatant forces in the continental United States.

Carl G. O'Berry retired from the U.S. Air Force as a Lieutenant General in August 1995. Until December 1998 he was vice president and director of planning and information technology for the Space and Systems Technology Group at Motorola, where he was responsible for Group-wide strategic and long-range planning and executive management of group information technology solutions and services. In addition, he was responsible for information technology architectures and roadmaps, new information technology business development, and leadership of information technology innovation and process reengineering. He was previously Deputy Chief of Staff for Command, Control, Communications & Computers, Headquarters, United States Air Force, a position from which he directed Air Force-wide information systems planning and policy development. Earlier in his Air Force career, he served as Commander of the Air Force Rome Air Development Center and as Joint Program Manager, World-Wide Military Command and Control System Information System. He also led the development and field testing of an airborne radar sensing/tracking system that was the forerunner of the Joint Surveillance and Target Attack Radar System. He has a master of science degree in systems management from the Air Force Institute of Technology and a bachelor of science degree in electrical engineering from New Mexico State University.

John H. Quilty is senior vice president and general manager, Washington C3 Center of the MITRE Corporation's C3I Federally Funded Research and Development Center. The Washington C3 Center supports the Army, Navy, Defense Information Systems Agency, Office of the Secretary of Defense, Office of the Joint Chiefs of Staff, and other members of the national security community. Mr. Quilty's current activities are focused on support of DOD initiatives and activities designed to achieve improved C4I support to joint operations. Previously, he assisted the general manager as vice president, Washington C3I Division, from 1986 to 1990. He is a member of the executive committee of the Armed Forces Communications and Electronics Association (AFCEA) board of directors and serves on the board of the annual NATO workshop addressing alliance issues following the end of the Cold War. He also serves as the Chair of the Military Communications Conference Board (IEEE/Armed Forces Communications and Electronics Association-sponsored). Mr. Quilty received a master of science degree in electrical engineering from Stanford University in 1962 and a bachelor of science degree in the same discipline from Princeton University in 1961.

Robert H. Reed is a director of the Lear Astronics Corporation, a company that produces flight control computers and associated software and develops unique applications of radar and other sensor technology. Previously, he was the executive director of the National Training Systems Association, a trade association of companies producing computer-based training systems, programs, and products. General Reed served with the U.S. Air Force during the period from 1953 through July 1988. His last military assignment was to SHAPE (NATO), Mons, Belgium, where he served as Chief of Staff. He held the rating of Command Pilot with more than 6700 flying hours, including 339 hours of combat flying. He obtained a B.A. in political science from Syracuse University and a graduate degree in public administration from George Washington University.

H. Gregory Tornatore is the program area manager for Defense Communications Programs at the Johns Hopkins University Applied Physics Laboratory (JHU/APL). His areas of expertise include military command, control and communications (C3), wide-area surveillance, over-the-horizon sensors and targeting, communications networks and architectures, high-frequency radar, and ionospheric propagation. His current responsibilities include overall management of a diverse set of programs sponsored by Army, Navy, Air Force, and selected DOD agencies that address operational and technical issues associated with National Command Authority connectivity to U.S. strategic forces; DOD satellite communications architecture development, control, and network management; tactical C3 systems vulnerability assessment; anti-jam and low-probability-of-intercept tactical radio systems; advanced phased-array antenna systems; and intelligence and information operations. Mr. Tornatore also chairs the Applied Physics Laboratory's Internal Research and Development Command and Control Thrust Area, responsible for the application of new technology to DOD C3 problems. Mr. Tornatore has been employed by JHU/APL since 1977 and has been a member of the Principal Professional Staff since 1980. Prior to joining JHU/APL, Mr. Tornatore was employed at the Electro-Physics Laboratory, ITT Avionics Division. Mr. Tornatore received a master of science degree in physics from the Pennsylvania State University in 1964 and a bachelor of science degree in physics from St. Francis College in 1961.